THE BRAIN IN MOTION

THE BRAIN
IN MOTION

From Microcircuits to Global Brain Function

STEN GRILLNER

The MIT Press
Cambridge, Massachusetts
London, England

The MIT Press would like to thank the anonymous peer reviewers who provided comments on drafts of this book. The generous work of academic experts is essential for establishing the authority and quality of our publications. We acknowledge with gratitude the contributions of these otherwise uncredited readers.

This book was set in Adobe Garamond and Berthold Akzidenz Grotesk by Westchester Publishing Services. Printed and bound in the United States of America.

Library of Congress Cataloging-in-Publication Data

Names: Grillner, Sten, 1941– author.
Title: The brain in motion : from microcircuits to global brain function / Sten Grillner.
Description: Cambridge, Massachusetts : The MIT Press, [2023] | Includes bibliographical
    references and index.
Identifiers: LCCN 2022054177 (print) | LCCN 2022054178 (ebook) |
    ISBN 9780262048200 (paperback) | ISBN 9780262375313 (epub) |
    ISBN 9780262375306 (pdf)
Subjects: LCSH: Locomotion. | Neurophysiology. | Comparative neurobiology. |
    Motor ability. | Movement disorders.
Classification: LCC QP301 .G685 2023 (print) | LCC QP301 (ebook) |
    DDC 612.7/6—dc23/eng/20230315
LC record available at https://lccn.loc.gov/2022054177
LC ebook record available at https://lccn.loc.gov/2022054178

10   9   8   7   6   5   4   3   2   1

Contents

Preface

I have written this book to convey my views on how the brain controls our movements, having worked on the design of microcircuits as well as global brain functions of the nervous system, from the spinal cord to the forebrain, using animal models, sometimes including humans. I will apply a combined motion and evolutionary perspective since most control systems are conserved, although new designs have been added along vertebrate evolution. Many of the views I hold today on how the brain operates have been sharpened or emerged from numerous enjoyable discussions with colleagues and collaborators worldwide and with many postdocs and PhD students, to whom I am of course indebted.

I begin the book by dealing with the evolution of the behavioral repertoire of vertebrates, followed by the function of the many microcircuits that underlie the basic parts of our motor repertoire and instinctive behavior, ranging from the hypothalamus to the spinal cord. This is followed by the midbrain circuits analyzing the egocentric world, and then the basal ganglia interacting with the cortex and cerebellum to achieve an integrated control of the various motor circuits and behavior in general. I am arguing that the forebrain is like the conductor of an orchestra, while the microcircuits for reaching, grasping, posture, locomotion, and numerous other patterns of behavior correspond to all the members of the orchestra. The conductor determines when each of them will be called into action.

With the broad scope that I chose, it follows that I will discuss some areas in greater detail than others due to my personal preferences, not necessarily because they are more important. I cite to a large extent reviews, and to some degree the original references as well, but will have neglected many, given the space limitations of this book, and ask for your understanding.

To some degree, this book reflects my scientific foci over an extended period. My early interest was related to the spinal organization of locomotion in mammals, and subsequently to understanding the intrinsic function of the central pattern generator (CPG). Next, I moved to the lamprey spinal cord (with Peter Wallén) and then extended our studies to the supraspinal control from the brainstem and the control of body orientation. From then on, my experimental work has often been combined with simulation as a complementary tool (with Anders Lansner and Jeanette Hellgren Kotaleski). After studying the visuomotor coordination of gaze mediated through the midbrain, we decided to explore the forebrain mechanisms engaged in the selection of behavior. To our surprise, we found that the detailed organization of the basal ganglia was conserved over 500 million years from the lamprey to mammals, and the sensory and motor areas in the mammalian cortex were also found in the lamprey. The blueprint of the vertebrate nervous system had thus already evolved when the evolutionary line leading to mammals became separate from that of the lamprey, although the number of neurons has increased by orders of magnitude, and thereby the potential for a more varied behavioral repertoire as well. This evolutionary perspective has been a rewarding focus for my work over the last several years.

In the process of writing this book, I have had the indispensable help of my colleague Dr. Brita Robertson, who has scrutinized and commented on all parts of the text and designed many of the figures; Professor Abdel El Manira, who has also read and commented on all the chapters as they were written, and Professors Jeanette Hellgren Kotaleski and Gilad Silberberg and Johanna Frost Nylén, a senior PhD student, who have given me very valuable feedback on some chapters. Dr. Lena Grillner, my wife, has provided an unimaginable amount of understanding and support of my priorities throughout my scientific journey, including the completion of this book.

My research has been funded each year since 1970 by the Swedish Research Council, by many grants from the European commission over many decades, and also by many other foundations.

1 THE VERTEBRATE MOTOR REPERTOIRE AND THE EVOLUTION OF THE BRAIN

To move things is all mankind can do—whether in whispering a syllable or in felling a forest.

—C. S. SHERRINGTON (1924)

1.1 INTRODUCTION

All living creatures need to interact with their environment, and even the most basic creatures have a set of innate motor circuits that can be called upon to feed, locomote, fight, or flee. In this chapter, I will show how the motor repertoire of vertebrates has evolved from that of protovertebrates to that of primates. I will consider all types of movements, whether they are the delicate finger movements of a pianist or whole-body movements such as locomotion and posture. I will avoid making any distinction between voluntary movements and basic motor patterns, since both walking and playing the piano are performed at will. I will also argue that the bases for practically all movements are innate circuits. Skilled movements often utilize innate components that can be programmed in a particular sequence. I have argued earlier that contrasting innate versus learned movements provide a false dichotomy (Grillner & Wallén, 2004).

An important message will be that we have a rich motor infrastructure extending from whole-body movements to reaching and independent finger movements in primates, and that these different building blocks can be recruited to design more complex movements. A skilled movement can be produced by combining distinct innate components in a well-timed sequence and by tinkering with the amplitude of the various components, whether the activity is playing tennis or writing a letter. These novel integrated motor patterns, which can be perfected through training, represent new skilled motor programs. They are put together by innate components from the motor infrastructure, and the learned parts comprise mostly specific timing and fine-tuning.

The neural circuits for generating the human movement repertoire reside mostly in the midbrain, brainstem, and spinal cord, while the forebrain determines when these circuits shall be recruited and the precise sequence of activating the innate motor components, sometimes called "motor primitives." The role of the cortical circuits and the basal ganglia together is thus to determine when the circuits for execution of movements shall become active. The analogy to an orchestra given in the preface and shown in figure 1.1 is relevant: the conductor (forebrain) determines what should be done and when, while each member of the orchestra (each motor microcircuit) performs his or her part, resulting in a well-coordinated symphony (skilled movement).

The orchestra – midbrain to spinal cord

The conductor – the forebrain

Figure 1.1
The orchestra and its conductor. The nervous system resembles in a sense an orchestra, each member of which has a special skill (e.g., playing the violin or the flute); but when and how the members play is under the control of the conductor. All the various motor programs or microcircuits of the nervous system are located in the midbrain-spinal cord.

I will start by giving a brief evolutionary perspective on the vertebrate motor repertoire, followed by the circuits for executions of movement, representing a motor infrastructure of innate motor circuits located mostly in the brainstem and spinal cord. Then comes a description of the intricate forebrain processing with close interaction between the cortex and basal ganglia, as well as considering the contribution of cerebellum.

1.2 VERTEBRATE MOTOR BEHAVIOR FROM LAMPREY TO HUMANS: OVERVIEW IN AN EVOLUTIONARY PERSPECTIVE

Nothing in biology makes sense except in the light of evolution.

—T. G. DOBZHANSKY

The oldest group of living vertebrates is the cyclostomes, represented by the lamprey, and the youngest groups comprise the birds and primates. Next, we discuss the range of motor behaviors that have evolved in the different classes of vertebrates (see figure 1.6 later in this chapter) in an evolutionary perspective. They can roughly be divided into the following groups:

- Whole-body movements, such as locomotion and posture, are represented in all classes of vertebrates, from swimming in fish and flying in birds to walking in tetrapods and bipeds. They are a precondition for escaping, freezing, foraging, and exploratory behaviors.
- Specific movements of the appendages, such as reaching and grasping, gradually evolve from amphibians to primates.
- Orofacial movements are used for ingesting food, manipulating objects, and in defense or attack. Included here are the associated motor patterns of chewing and swallowing.
- Respiratory movements are required to inhale oxygen and exhale carbon dioxide through the gills or lungs. Volitional control of the airflow is critical for vocalization and also depends on the larynx and the dynamic shape of the oral cavity, whether in frog calls, birdsong, or human speech.
- Various species display characteristic courtship behaviors, which are different for males and females, as well as specific motor patterns underlying

spawning or mating. Birds and mammals handle their young through characteristic maternal (and, in some species, paternal) behaviors.

- Many birds and mammals express different forms of emotions through vocalization and facial or bodily motor patterns.
- All classes of vertebrates have similarly organized eye movements and display both orienting movements, toward an object of interest, or evasive movements, to avoid collision or confrontation with objects in the surrounding environment.

1.2.1 Whole-Body Movements

1.2.1.1 Posture

During locomotion, most vertebrates, from fish to birds and mammals, have the body oriented with the back in a dorsal position, and a few bipedal species such as humans have the trunk in a vertical position. The vestibular sense organs play a dominant role in the control of body position in fish, as in other vertebrate groups, in which proprioceptive reflexes from the limbs also have made significant contributions for maintaining the orientation of the body. The projection of the center of gravity needs to be kept dynamically between the various points of support as the limbs move. There is a seamless transition from locomotion to posture, as elegantly formulated by Sherrington's statement that "posture follows movement like a shadow." (see Grillner & El Manira, 2020). Most likely, the postural control system is integrated with the locomotor system during ongoing locomotion, and when the movements stop, the postural system continues to operate independently in the absence of locomotor movement.

At rest, the body can voluntarily assume a wide variety of positions, such as standing, sitting, lying, or with the head bent forward. To provide the appropriate neural commands requires integrated information about the positions of all the parts of the body in relation to each other, which is provided mainly by the proprioceptive system. Similarly important is information regarding the orientation of each body part in relation to gravity. This information, partly processed dynamically in the spinal cord and cerebellum, is available continuously to the central nervous system and is sometimes referred to as the "body scheme." The position of the body at any given point of time serves as a platform for other movements (Gray, 1968), as when reaching out to grab a ball.

1.2.1.2 Locomotion

The lamprey has a long, slender, eel-like body without any paired fins, and it swims with undulatory swimming movements propagated along the body (figure 1.2a). These movements are steered by adding excitation to the motoneurons on one or the other side of the body. The trunk movements are controlled by the segmental motoneurons, located along the spinal cord in a medial motor column (MMC). In addition, vertebrates with limbs or paired fins have a lateral motor column (LMC) at both the cervical and lumbar levels, which innervates the forelimbs and hindlimbs or the corresponding fins.

Next in line, the sharks (elasmobranchs) represent the second-oldest group of vertebrates and also swim with undulatory waves propagated along the body. They have evolved paired pectoral and pelvic fins that are controlled by the motoneurons of the newly evolved LMC. The large pectoral fins in the

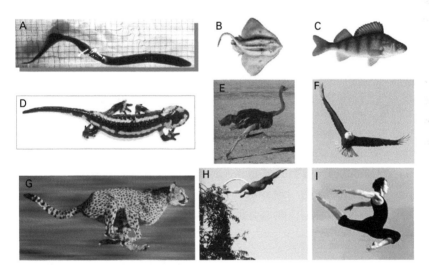

Figure 1.2
Different forms of locomotion among vertebrates. *A.* Lamprey. *B.* Elasmobranch (skate). *C.* Teleost (perch). *D.* Amphibian (salamander). *E.* Ostrich (walking). *F.* Eagle (flying). *G.* Cheetah (the fastest mammal with regard to locomotion). *H.* Primate (arboreal locomotion). *I.* Variation of human locomotion.

shark are used very efficiently for steering during locomotion. In the related skate, the pectoral fins have been further enlarged and take part in the propulsion, with a dorsoventral traveling wave from rostral to caudal (figure 1.2b). A remarkable and unexpected finding (Jung et al., 2018) was that the pelvic fins of the ray can induce alternating movements of the fins in a leglike fashion that result in forward motion, when the ray rests on the bottom of the sea and is not actively swimming.

In the teleosts (bony fish), the pectoral fins have gained further independence and can be moved independently (figure 1.2c), such as when the fish is positioning itself to eat snails on a rock. The fins can also contribute to locomotion at slow speeds, while at higher speeds, the undulatory trunk movements dominate and the pectoral fins are folded to the side of the body. This very large group of species (numbering around 20,000) displays many variations and body shapes. Some fish, such as the mudskipper (periophtalmus lineage) in the Goby family, have their pectoral fins directed ventrally and can be used as propulsive appendages. The mudskipper is thus able to move on land for shorter distances than in the water, while the body is kept stiff.

Amphibians constitute the first tetrapod group that evolved forelimbs and hindlimbs for locomotion with the same skeletal parts as in mammals, including "fingers/digits," and the same muscle groups in the limbs and trunk. The salamander group swims with fishlike, undulatory movements with the limbs kept close to the body. At slower speeds, they walk either on the bottom of ponds or on the ground with an alternating gait and with the limbs held in a lateral direction that ascertains a more stable control of body posture (figure 1.2d). The other amphibian group, the anurans, comprises frogs and toads, which have a more compact and square body shape. Frogs use mostly a hopping gait, with the hindlimbs extended in phase (synchronous) with the propulsive phase, but they can also alternate. While the animal is in the air, the hindlimbs are flexed while the forelimbs become extended as the frog approaches the ground. This means that at the amphibian stage, the basic modes of "tetrapod coordination" had already evolved with alternating gaits as in walking or trotting, or "in phase coordination," as during hopping or mammalian galloping.

Reptile locomotion for many groups, such as crocodiles and lizards, is similar to that of the salamander, but turtles have stiff shells and use their limbs only for walking and swimming. Finally, snakes have lost their appendages and display a variety of undulatory movements; for some, such as eels or lampreys, it is swimming, while others display more specialized forms, like the sideways movements of the sidewinder (Gray, 1968).

The bird lineage represents a further extension of the reptilian sauropsids (dinosaurs) group, and their forelimbs have been restructured to wings, but with a similar basic set of skeletal parts and muscles. Birds have sacrificed the reaching and grasping ability of the forelimbs for being able to explore the world in three dimensions, like the eagle shown in figure 1.2f. The wings are used in phase during flight rather than in alternation, as in walking. The hindlimbs are mostly used for bipedal walking (the ostrich in figure 1.2e) or hopping (the magpie), or for swimming (ducks or gulls, in this case generally with webbed feet). The legs can also be used to secure the location of pieces of food on the ground, while the bill is for grasping. Some birds, such as ospreys or eagles, can also use the hindlimbs during hunting to seize a fish or another prey with their feet and transport it to a safe place. This prehensile ability of the hindlimbs is a unique function that is normally a task for the forelimbs in other vertebrate groups.

The basic motor patterns of standing, flying, and landing are innate. I once observed a young nestling of a blackbird that had its nest on the windowsill on my balcony. The young blackbird, ready to leave the nest, was standing on the rim of the nest, and then it flew to the railing of the balcony and grasped it. A minute later, it flew down some five meters to the ground and, just before landing, flapped its wings to stop the downward momentum and land almost perfectly. Given that this was the fledgling's very first trial, it was an impressive event to observe. At the same time, this shows that the entire sequence of motor patterns used in an appropriate sequence (flying, grasping the rail, flying again, and landing) must be genetically coded as well as the basic aspects of visuomotor coordination, since they were able to handle this difficult task on the very first trial.

In mammals, the alternating gaits are walking and trotting, and in some species pacing. In the latter case, the forelimbs and hindlimbs on the same side are synchronous rather than alternating, as in a trot. Most species also display in phase movements such as in the forms of galloping and bounding as demonstrated by the cheetah in figure 1.2g that can actually reach over 100 km/h during a sprint and represents the fastest running mammal. Sea-living species, such as whales and seals, move mainly through dorsoventral movements of the body and the appendages have been reduced in size and are mostly used for steering.

Arboreal mammals, such as the squirrel, jump from branch to branch and have developed an ability to grasp very efficiently as they approach a new target. Primates, notably the orangutangs, have evolved the capacity to use the forelimbs independently as they swing from one branch of a tree to another (figure 1.2h). This means that a primate must have a precise estimate of the location of the new target, elicit an appropriate reaching command, and upon reaching the target, be able to firmly grasp the new branch. At this point, it needs to carry the weight of the entire body on one arm, swing forward to a new target, and use the other arm (or, more precisely, the hand) for grasping. Most primates, such as baboons, use all four limbs for locomotion, but many monkeys and apes are bipedal for part of the time (Higurashi et al., 2019). This has the advantage of freeing the forelimbs for grasping objects, such as palatable fruits, or (for some apes) using stones as tools. Having the body upright, of course, means a better overview as animals move around in a landscape that offers many opportunities, but also threats. The bipedal locomotion of nonhuman primates is different from that of humans, in that the knees are bent during the support phase in the tetrapod gait, whereas in the human walk, the legs are kept straight. During running, however, the limbs remain bent and springy during the support phase (Grillner et al., 1979). Humans can combine locomotion with a variety of impressive motions, such as those of the ballet dancer in figure 1.2i.

The basic pattern of coordination of the hindlimbs, as viewed by electromyographical activity in the leg muscles, are similar in different mammals (rodents to primates), reflecting the four phases of each step cycle (support, liftoff, forward motion of the leg, and touchdown). This basic pattern of coordination is also found in birds such as guinea fowls (Dominici et al., 2011). This implies that the four-phase motor pattern existed in the last common ancestor, which would be an early reptile, or possibly an amphibian. The trunk and forelimb motor patterns are thus conserved, even though the efficacy of the locomotor movements varies remarkably from the impressive run of a cheetah to the lethargic movements of a turtle. This is likely accounted for by a conserved organization of the locomotor networks at the level of the spinal cord.

1.2.1.3   Flight, freeze, or fight

An important priority for all vertebrates is to be able to react to an impending threat by fleeing through an almost instantaneous activation of the locomotor system. Many vertebrates instead react to a potential threat by becoming inactive—that is, to freeze and remain still in a given posture. The reason for this is that most predators detect prey as they move, and therefore the "playing dead" reaction can be lifesaving. The third possibility is to attack and fight, which requires not only the locomotor system, but also a flexible, well-adapted body and jaws.

1.2.2   Reaching, Grasping, and Accurate Foot Placement during Locomotion

The paired pectoral fins in fish can be used independent of each other for positioning themselves, such as in relation to pieces of food. The proximal parts of the fins evolve into the forelimbs of amphibians with the basic skeletal design of a shoulder, upper and lower forelimb, hand bones and digits, and the related muscle groups. This design is used throughout vertebrate phylogeny, although different adaptations have taken place, such as in hoofed animals.

Amphibians and reptiles clearly steer their locomotor movements with the help of the forelimbs and use their distal, fingerlike appendages for grasping various structures in the context of locomotion, as when they move along the stem of a tree and reach out for support. There is a large array of specializations of the "hand," particularly in reptiles, to enhance friction and facilitate movements on leaning surfaces. This can take many forms, such as pads on digits.

In many species, the forelimbs are also used for independent reaching and grasping, such as during feeding and prey capturing in many frog species (Gray et al., 1997). Frogs can reach out with a forelimb to grasp prey (e.g., an insect) and bring it to the mouth. The digits of the frog's hand are at first extended

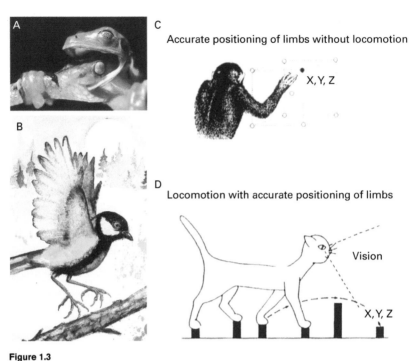

A

B

C

Accurate positioning of limbs without locomotion

X, Y, Z

D

Locomotion with accurate positioning of limbs

Vision

X, Y, Z

Figure 1.3

Reaching is present from amphibians to primates. *A.* A frog grasping to remove an attacker (coutesy of David Maitland). *B.* The passerine, prepared to grasp around the branch of a tree, while landing. *C.* A primate pointing toward a target. *D.* Precision walking of an animal requires placing the limbs on specific targets and the same corticospinal neurons are used for reaching and during locomotion.

and then contracted around the object as it grasps (figure 1.3a). Frogs also use the forelimbs to push food into the mouth.

Many mammals can reach for specific targets, as when eating, when the food can be secured by one hand or foot while the jaws grab pieces of the food item. This strategy is typically found in dogs. Rodents and cats use more advanced movements and have the capacity to grasp small pieces of food such as seeds with their paws and bring them to the mouth. The paw must first be oriented toward the location of the food and then be reshaped. The paw is held comparatively open for grasping larger objects, but with the digits more contracted if the object is small (Whishaw et al., 1992; Whishaw & Coles, 1996; Iwaniuk & Whishaw, 2000). The successive stages in the grasping behavior are remarkably similar in rodents and humans (Sacrey et al., 2009), which suggests that the neural machinery for reaching, and grasping is conserved, probably based on the grasping circuitry already present in frogs. Mice can even peel off the capsule of a seed while holding it in the paw, using the mouth/teeth for handling the capsule of the seed—indeed a sophisticated case of fine motor control. They can also grasp elongated objects such as a piece of pasta. The neural bases of reaching and grasping circuitry have been explored in some detail (Ruder et al., 2021; Esposito et al., 2014).

1.2.2.1 Many primates have the ability to move each finger independently

Most animals use the paw/hand to contract or extend the digits synchronously so that it functions as a unit. A further development in some primates is independent control of individual digits, which is required when an object must be retrieved from a small hole, for instance. The control of individual fingers requires an intact corticospinal system. This control is lost after lesions of the corticospinal tract (Porter, 1987), but it gradually recovers, although not at the same level. Finally, the human hand is an outstanding, versatile tool that can be used in numerous configurations with the different fingers, such as when carrying heavy objects, creating novel pieces of art (like that of a silversmith), or playing the piano or the violin (see chapter 5). This all depends on our ability to use our fingers in amazing combinations. Our ability to move the fingers independently is innate and typical of our species, but we can learn to use our fingers in a variety of contexts and adapt the movements to new situations like tying shoelaces (Dhawale et al., 2017, 2021).

1.2.2.2 Birds reach out and grasp with their hindlimbs

Many birds, particularly passerines, can rest while sitting on a small twig of a bush or tree. This depends entirely on their ability to grasp the small twig with their feet (hindfeet) when landing, based on previous visual information used to orient the feet appropriately (see figure 1.3b). In addition, they need to have an astounding equilibrium control as they maintain body orientation as the branch that they are sitting on is moving in the wind, which requires exquisite vestibular and proprioceptive control. Birds can also use one limb to secure the position of a piece of food on the ground while using the beak to subdivide the food into palatable parts. As mentioned earlier, eagles and ospreys use their feet to capture prey, such as a fish or a rabbit, and then carry the prey while flying to their nest or some other location.

1.2.2.3 The role of reaching circuits during locomotion

The forelimbs are used for steering locomotor movements by rotating the limbs in the desired direction before they are placed on the ground, together with orienting the body in the same direction. In addition, the forelimbs are used for an accurate placement of the feet during each step on uneven terrain. This is like performing a reaching movement toward a specific position, but with the added complication of the ongoing locomotor movements (Georgopoulos & Grillner, 1989). That the same neural machinery is actually used during reaching and locomotion is shown by the fact that the very same corticospinal neurons are activated both in a reaching task and during precision walking in the cat (figure 1.3c,d; also see Yakovenko & Drew, 2015). The same neural control system can thus be used independently for reaching toward different targets in the environment or in combination with locomotion for accurate foot placement with each step. It can even be argued that the origin of reaching movements is related to the requirement to place the limbs at an optimal point of support during locomotion.

1.2.2.4 Rhythmic scratch reflexes or grooming movements

In most tetrapods, it is important to keep the body clean and remove irritants from the skin. In amphibians, one hindlimb is used to wipe away irritants on the same side of the body. In mammals, as well as in birds and reptiles, a scratch reflex occurs which is directed to remove irritants on the skin. This requires rhythmic movements of the leg, and in addition, the paw is directed to the specific location of the irritant on the body. The leg performs rhythmic movements resembling those of locomotion, but the hip joint is held in a position so that the paw can make contact with the affected skin area. This seemingly complex process, requiring a body plan, can be performed by the spinal cord circuits without any contribution from higher centers (Stein, 2008; Fukson et al., 1980; Sherrington, 1906). Either one or two forelimbs perform grooming movements that are also rhythmic and directed to cleaning particular areas on the head and face. These movements are coordinated at the brainstem–spinal cord level (Kalueff et al., 2016). In some species, such as the horse, not all parts of the body can be reached by a limb, and the animal may then rub the body against a tree, for instance, to remove the irritant.

1.2.3 Orofacial Movements: A Main Source for Interaction with the Surrounding World

Although most vertebrates interact with the environment by using their mouths and jaws, the neural control of this important motor behavior has been much less studied than control of the limbs. The oldest vertebrate group, the cyclostomes, lack jaws but have a circular mouth, which is used to efficiently suck onto their prey, while the rough tongue is used to penetrate the skin of the prey. Over an extended time, they suck the body fluids from the victim, which can become seriously weakened or even die.

All the other vertebrates have jaws that are used not only for ingesting food, but also for interacting with the surroundings and in defense, attack, or play. For most species, the orofacial control system is the most important neural machinery for exploring and interacting with the surrounding world. In monkeys, apes, humans, and some other species, the arms and hands have instead at least partially taken over this role.

For feeding, most animals grab large or small pieces of food and simply swallow, whether they be sharks, crocodiles, or seagulls. Further parcellating the food by chewing before swallowing is mainly a mammalian development, and it is coordinated by networks in the brainstem (Dellow & Lund, 1971). This entails opening and closing the jaws, as well as moving the tongue in the opening phase to relocate the food in the mouth. When the food is brought to the posterior parts of the pharynx, a swallowing network becomes activated that brings the food down to the ventricle by a programmed sequence of muscle actions (Jean, 1990, 2001).

Mammals also use the jaws for carrying food or their young. A dog may carry some object back to its owner, and the jaws also can be used in hunting, as when a cheetah attacks an antelope. The latter behavior is a whole-body engagement, but the jaws play the central role in killing the prey, as also done when animals fight for their life. On the other hand, when pups are playing, they abstain from severe bites but simulate fighting and learn to perfect their movement repertoire.

1.2.4 Breathing, Vocalization (Including Birdsong and Human Speech), and Expression of Emotions

1.2.4.1 Respiration

All vertebrates breathe with rhythmic movements to produce water flow over the gills or airflow to and from the lungs. The rhythmic movements are in both cases produced by brainstem networks of a similar design (Del Negro et al., 2018; Cinelli et al., 2013, 2016; Missaghi et al., 2016), although the downstream respiratory structures may be arranged in a radically different way in various animal groups. The respiratory regulation of breathing frequency and depth of breath is regulated by chemoreceptors that sense the carbon dioxide and oxygen levels in the blood.

1.2.4.2 Vocalization, including speech

The airflow of the lungs is also used for vocalization, along with a modulation through the larynx and the dynamic shape of the oral cavity. A set of innate warning calls are used by most birds and mammals to signal a possible threat. Neural circuits in the midbrain, including the periaqueductal gray (PAG), are responsible for eliciting the commands for vocalizations of this type. The crying of a newborn baby belongs to this category, as does other sounds, such as the barking of a dog or the purring of a cat.

Birdsong is a more sophisticated form of vocalization that is often species-specific and produced by neural circuits in the forebrain. In most songbirds, it is seen as depending on some sort of vaguely defined innate neural template, which is different for each species. The song is reinforced by the young having heard it in early life and have it crystallized when it starts to sing later in life. However, the motor sequences can be modified extensively through learning by exposing the bird to modifications of the original song. Some birds, such as the parrot and the mockingbird, have the unique ability to learn through imitation, including the calls of a variety of other birds, traffic sounds, and human sounds.

The same principles are used to some degree in the most unique human behavior, speech, which allows us to communicate about past, present, and future events (see also chapter 5). As children, we learn the words and logic of our own native language, as well as the various sounds that are used. As the brain gradually matures, more advanced processes evolve as cognition. It all, however, depends on the ability to dynamically control the respiratory flow, the dynamic preshaping of the oral cavity, and the position of the tongue. This astounding ability to develop language has ultimately allowed the development of civilizations, together with the possibility of transferring knowledge through writing. The latter invention, written language, evolved only late in the short history of our species.

1.2.4.3 Expression of emotions

Most animals, from fish to mammals, use facial expressions or body postures with vocalizations to express anxiety, aggression, satisfaction, maternal behavior, and courtship behavior. In 1872, Charles Darwin published *The Expression of the Emotions in Man and Animals*, one of his many remarkable books. He could show that the expression of human emotions such as smiling, laughing, crying, anger, and worry is innate and expressed in a similar way in different parts of the world. Figure 1.4 shows reproductions of some of the original plates of Darwin, showing a smiling girl, a weeping child, a worried boy, a woman expressing disapproval, and finally an angry toddler. One must recall that using photography in 1872 at the time was a high-tech enterprise. The sculpture of the angry boy, made by Gustav Vigeland, is displayed in Oslo. Darwin pointed out that, for instance, children born blind smile as beautifully as children who can see, and therefore the neural circuits that generate the smile must be innate. He also discussed the role of the facial muscles that could be activated to elicit the facial expressions underlying the various emotions. He concluded that each type of emotion was based on a combination of active facial muscles. He actually reported the effect of selective stimulation of the different facial muscles (e.g., the forehead, eyebrows, and shape of the mouth) and how they contribute to the expression of the different emotions (i.e., in a sense, a set of motor programs).

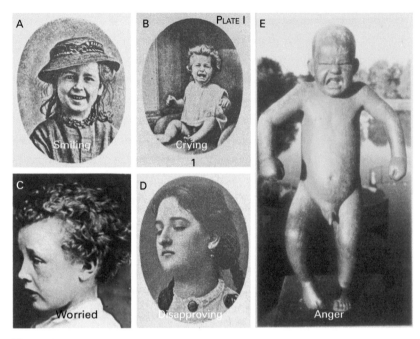

Figure 1.4
The expression of different forms of emotions in humans depends on specific motor programs. Some of Darwin's (1872) original illustrations of the expression of emotions in humans (a girl smiling, a child weeping, a worried expression of af a boy, and a woman expressing disapproval) are shown in (A)–(D). Panel (E) shows a sculpture of a boy displaying anger made by Gustav Vigeland.

Humans are now considered to have seven characteristic facial expressions (Ekman et al., 1987). Our motor programs to express emotions are innate, but what we laugh at or react to with disgust is culturally bound. Darwin also noted that other mammals such as dogs display signs of pleasure when meeting individuals that they know, or anger, or submissive behavior when they have misbehaved.

1.2.5 Eye, Orienting, and Avoidance Movements

Most vertebrates depend on vision, but to varying degrees. Eyes with well-developed retinas exist in the lamprey and all later-evolved classes of vertebrates. The visual information is processed first in the retina and then transferred in a retinotopic fashion to the midbrain roof (tectum, called the "superior colliculus [SC]" in mammals) and in parallel to the visual area in pallium in the lamprey, and for object recognition to the visual cortex in mammals. The retinotopic map is located superficially in the tectum/SC. In register, but deeper, there is a motor map, which can elicit eye movements toward the part of the visual field that has been activated by the overlying retinotopic area or orienting movements of the head and body. The control of eye movements is through the same groups of eye muscles in the lamprey and other vertebrates and controlled by the same cranial motor nuclei (third, fourth, and sixth).

To be able to maintain stable vision, the gaze is controlled through vestibulo-ocular reflexes producing compensatory eye movements as the head moves and optokinetic reflexes elicited when an object is drifting over the retina. In addition to eye movements, orienting movements of head and body are elicited. If a large object appears ahead, a different set of tectal neurons become activated to avoid collision and elicit an avoidance movement. These mechanisms, including nystagmus, exist in both the lamprey and mammals and most likely represents a basic neural design feature that is required for an optimal usage of vision during movements in all vertebrates.

A different class of eye movements includes those used to track a visual object, as when watching a tennis ball moving back and forth. In mammals, this is processed in the cortex. It is not yet clear which vertebrate groups possess this capability. The tracking movements require much more time for processing than the vestibulo-ocular compensatory movements. Thus, the basic features of control of the gaze developed very early in vertebrate evolution.

Animals that often move in dark habitats, such as rats, mice, and the cat, have developed complementary tactile sensors, the whiskers, which are very sensitive and respond to the slightest contact at the far end of each whisker. They are used extensively by rodents to maneuver in dark passages and rely on a complex processing in the barrel cortex and in other parts of the nervous system.

1.2.6 Innate Motor Programs during Ontogeny: Maturation and Plasticity

When a motor program starts to operate in the very young, it is usually far from perfect and reaches completion as the animal grows. The motor program may also become optimized as it is used—practice makes perfect.

Many mammals, such as cats and dogs, play extensively during the first few months of life with their littermates. Play no doubt provides a very versatile training and exposure to a variety of movements and makes their movement repertoire perfected.

The human baby is born quite immature and goes through stages of maturation during the first year (figure 1.5). When it is born, it can breathe, be fed, and cry, and a few other faculties are developed. With each month, the baby will successively be able to balance the head, to sit and stand without support, to crawl, and finally to walk a few steps. It is important to realize that this is essentially a biological maturation process, and the child is not "learning to walk," as the common expression says.

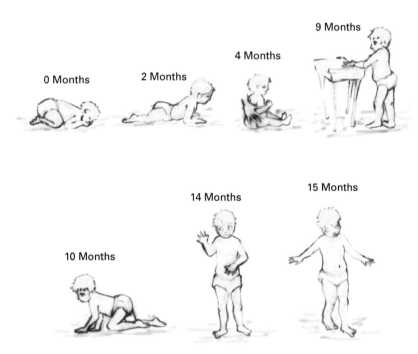

Figure 1.5
The progressive maturation of the motor development of the human infant and young child. At two months after birth, a child can lift its head; at four months, it can sit up with support. Subsequently, it can stand with support, crawl, stand without support, and finally walk. The approximate time at which a child can perform those motor tasks is indicated above each image. The variability in the maturation process is substantial among different children.

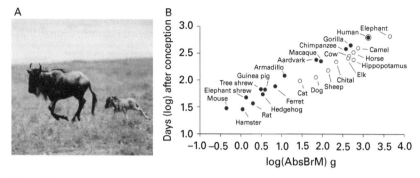

Figure 1.6
Mammals differ in their degree of maturation at time of birth. *A.* One extreme is represented by the antelope (wildebeest), shown here with a young calf galloping behind the mother, ten minutes after delivery. *B.* The adult brain weight of different mammals plotted versus the time to onset of walking calculated from day of conception (log-log scale). Note the linear correlation between brain weight and the time to start walking. Some species are very immature at birth, such as the human baby, and others very mature, such as the deer/elk, and yet they fall on the same regression line. Solid symbols represent species that can assume a plantigrade hindlimb standing position; open symbols represent species that cannot do so. Modified from Garwicz et al. (2009).

Other mammals, such as hoofed animals, are born very mature. A few minutes after birth, some antelopes are not just able to stand but also run, such as the calf of the wildebeest (antelope), which must be able to respond to the impending threat from predators by running away (see figure 1.6a). Likewise, some birds can stand and walk upon hatching and even pick up grains off the ground.

Why do we have this marked difference? Garwicz et al. (2009) provided an unexpected answer to this question, as they showed that there is a linear relation between brain weight and the time from conception until when an animal starts to locomote—from mice to humans and elephants. The latter have the largest brains (figure 1.6b). Locomotion is a complex task, considering many aspects that need to be integrated—locomotion, posture, and visuomotor coordination (to mention just a few), and the difficulty must be expected to increase with the weight of the animal. It is, of course, to be expected that it takes a much longer time to form the complex brain of a human than the tiny brain of a mouse.

Most animals grow gradually to a weight many times greater than that of the young pups or larvae. This means that the neural circuitry must adapt in a remarkable way to be able to control the increasing mass that it has to steer in actions such as reaching or locomotion—consider, for instance, the weight of a newborn compared to an adult elephant. The entire control systems for posture and locomotion need to be recalibrated as an animal grows. In another example, the locomotor network of the larval zebrafish becomes somewhat modified, and some neurons take on a new role on the way to becoming an adult, during which process the weight and size has increased many times. The maximal swimming frequency is high for the small larvae and much lower for the adult, although the speed of swimming can be much faster in the latter (see chapter 2).

As an animal grows, remarkable changes need to take place in the nervous system. This concerns the efficacy of the synaptic transmission, neuronal properties, number of neurons, and even the size of the nervous system. The latter will increase, and the length of many axons will likewise. The size and number of muscle fibers and the entire body will increase as well. Some vertebrates continue to grow throughout life, whereas others reach their permanent weight in adulthood.

1.2.7 Conclusions

Many aspects of vertebrate motor behavior have their origin early in vertebrate phylogeny, and all display escape reactions and freezing. As the limbs evolve, they can be used for steering, and, already in amphibians, they are also used for reaching and grasping, although the precision and versatility increase gradually in mammals. Primates add the ability of independent finger movements, which allow fine manipulation and enable humans to play the piano. Control of the eyes is conserved from the lamprey to primates, as well as the innervation pattern of the eye muscles and the control of orienting and evasive movements. The latter are critical when moving around in a complex terrain in which you need to avoid bumping into obstacles, and the orienting movements are needed to point the eye, head, and body toward a point of interest, as during foraging.

Table 1.1
Classes of vertebrates and patterns of coordination

	Locomotion (Undulatory)	Locomotion (Limbs)	Reaching	Grasping	Eye/Orienting Movements
Lampreys	+				+
Elasmobranchs	+	+			+
Teleosts	+	(+)			+
Amphibians	+	+	+	+	+
Reptiles	+	+	+	+	+
Birds		+	+	+	+
Mammals	Whales	+	+	+	+

Note: Among mammals, whales and related groups use dorsoventral undulatory swimming, and teleosts have paired fins but use them mostly in relation to steering and for positioning themselves (shown by the parentheses).

1.3 THE BASIC BUILDING BLOCKS OF BEHAVIOR: MOTOR PROGRAMS AND THEIR SELECTION—OVERVIEW

1.3.1 Execution of Movements: A Palette of CPG Networks and Motor Centers from Midbrain to Spinal Cord

As mentioned previously, all vertebrates possess a set of innate motor programs for locomotion, postural adjustments, respiration, swallowing, eye movements, and many more actions that form a motor infrastructure for execution of the basic motor repertoire of a species. These motor programs or central pattern generator (CPG) networks reside in the midbrain, brainstem, and spinal cord (figure 1.7). They can be recruited when needed by forebrain circuits.

Definitions:
- The term "CPG" is used for a circuit that generates a specific pattern of behavior, such as locomotion, respiration, or chewing, which are rhythmic, but also those that generate motor sequences such as swallowing or coughing.
- The terms "circuits" and "networks" are broader and more general and will include CPGs.

- The term "microcircuits" is mostly used to refer to parts of an integrated circuit, such as a part of the CPG or part of the hand-to-mouth synergy.
- Motor primitives overlaps with CPGs, but they can also be used for parts of a CPG, such as the flexor synergy during locomotion or the microcircuit generating saccadic eye movement.

Classic experiments from the Sherringtonian era and later have shown that well-coordinated movements can still be elicited in mammals in which the forebrain had been inactivated (decerebrate preparations). These animals, with only the midbrain and structures further down intact, can breathe and adapt their breathing to metabolic demands. If food is put in their mouths, they swallow and sometimes perform locomotor movements spontaneously or if stimulated in a specific midbrain locomotor command center, the mesencephalic locomotor region (MLR). Eye and orienting movements can be elicited by stimulation at different sites in the midbrain roof, such as the superior colliculus (SC). Figure 1.7 shows examples of the innate motor programs mostly available in both mammals and the lamprey. Thus, the spinal cord contains the networks coordinating locomotion, but they become turned on from supraspinal command centers. These circuits or CPGs determine when each group of motoneurons is activated in each step or swim cycle. The spinal cord also contains the circuitry for protective reflexes like the flexion reflex, which elicits a withdrawal of the limb from a harmful place or the scratch or wiping reflex, which birds, mammals, and amphibians use to remove irritants from the skin. The CPGs for respiration and chewing are located in the brainstem. The control of body orientation depends on vestibular processing at the brainstem level and sensory input from the limbs.

The many innate circuits, CPGs, and other networks form a palette of options for the forebrain to select among. In mammals, preformed neural circuits are thus available for eye movements, reaching, chewing, swallowing, respiration, locomotion, posture, protective reflexes, and many more actions in mammals. There is a similar organization in the lamprey, but circuits for the control of limbs are lacking, for obvious reasons (figure 1.7). These circuits provide the neural basis of the standard behavioral repertoire of any species. Each CPG provides a standard format for its motor pattern.

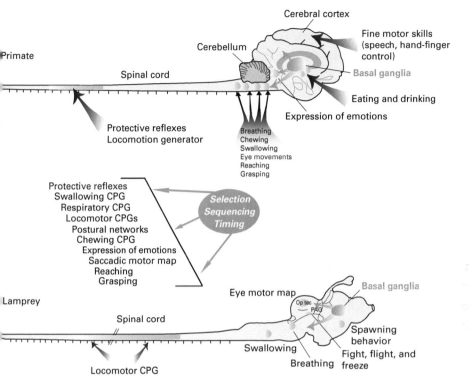

Figure 1.7

Common motor infrastructure from lampreys to primates. As indicated, the primate and lamprey brains contain the neuronal networks (CPGs) controlling different basic aspects of motor behavior. The networks are located in the spinal cord (e.g., locomotion), brainstem (e.g., swallowing, breathing), and midbrain (eye, orienting/evasive movements). The basal ganglia are similarly organized in the lampreys and primates and play a crucial role in the selection of motor behaviors. In primates as well as other mammals, the cerebral cortex contributes to the control of fine motor skills. Abbreviations: PAG, periaqueductal gray; Op tec, optic tectum. From Grillner and El Manira (2020).

In locomotion, for instance, a sequential activation of the various muscles results in a support phase of the limb, followed by a liftoff phase, a forward movement of the limb during the swing phase, and finally a placement of the foot on the ground. However, under natural conditions, each individual step will vary due to the terrain, and therefore the CPG activity needs to be able to adapt to these perturbations. This is handled by sensory input, signaling the load on the limb, and the position of proximal and distal joints. In addition, predicted perturbations such as upcoming obstacles can be handled by visuomotor signals from the cortex and other structures that become integrated with the CPG commands in the spinal cord.

1.3.2 Selection of Motor Programs a Forebrain Affair: Basal Ganglia and Cortex

For the selection of behaviors, several key forebrain structures are engaged, in particular the basal ganglia, including the dopamine system, cortex, and thalamus. They determine if a downstream motor program is to be recruited or if it is time to terminate an action pattern.

The CPGs in the midbrain, brainstem, and spinal cord are directly or indirectly under a tonic inhibitory control from the basal ganglia output nuclei, which is the substantia nigra pars reticulata (SNr) and globus pallidus interna (GPi). A recent study in the mouse has shown that the SNr targets no fewer than forty-two downstream target structures in the midbrain and brainstem (McElvain et al., 2021) and thus can exert control over them. The rationale for this is presumably to ascertain that the CPGs should remain inactive until specifically called into action. By selectively inhibiting a group of SNr/GPi neurons involved in the control of a specific motor program, the downstream center will be disinhibited and allowed to be recruited into action. The disinhibition due to the inhibition of neurons in the SNr/GPi can be complemented by excitation from other sources such as the cortex.

The basal ganglia thus play a central role in the process of selecting which action is to be initiated (for more, see chapter 4 and figure 1.10a). The striatum, the input structure, is the main processing center and will determine the level of activity in the SNr/GPi. Activation of the striatum is primarily driven by excitatory glutamatergic input from the various parts of the cortex, each affecting different compartments of striatum—a topical organization. The input to the striatum from the nuclei in the thalamus is almost as important. These nuclei, in turn, receive input from a variety of subcortical structures and the cerebellum. Another important factor is the dopamine neurons in the substantia nigra pars compacta (SNc) and the adjacent ventral tegmental area (VTA). Dopamine neurons are tonically active at rest and are further activated by salient stimuli in the surroundings and in reward situations, and conversely, they are depressed if an expected reward is not received.

The motor cortex projects heavily to the somatomotor part of the striatum (dorsolateral striatum in rodents) and directly to downstream centers in the midbrain, brainstem, and spinal cord. It is thus likely that when a downstream motor center is activated, there will be a combined disinhibition through the SNr and basal ganglia and direct excitation from the cortex (see chapters 4 and 5). The dorsomedial parts of the striatum receive input from the prefrontal and limbic areas and the ventral striatum from the hippocampus and limbic areas and are thought to process information of a more cognitive and emotional nature.

1.3.3   The Role of the Cortex: Far from Straightforward

The role that the cortex plays in the initiation of behavior is ambiguous and varies most likely under different conditions. Arber and Costa (2018) considered that the cortex may express a wish, while the basal ganglia would determine whether that wish would be fulfilled (figure 1.8). The Ölveczky laboratory (Kawai et al., 2015) showed that a learned behavior like a double pedal press with a fixed time interval could be produced with the same precision on the first trial after removal of the frontal motor areas, and thus the motor cortex is not needed for the initiation and execution of this learned motor sequence. However, the motor cortex was required during the learning period. There is thus an intricate interaction between the two structures.

In mammals, such as rodents, rabbits, and cats, removal or inactivation of the neocortex in young animals, leaving the rest of the brain intact, has surprisingly little effect on the standard motor repertoire. Such a decorticate animal will walk around, perform exploratory movements, search for food and water, eat, and go through phases of sleep and active behavior (Bjursten et al., 1976). They can live for years in a laboratory environment. This means that the process of initiation of different patterns of behavior can be handled by the basal ganglia, with input from the thalamus but without input from the cortex.

However, dynamic tasks requiring visual interaction with moving objects or fine manipulation of objects will most likely require the cortex. Also,

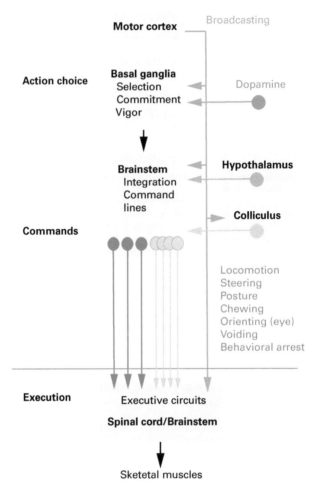

Figure 1.8
The contribution of various parts of the nervous system to the control of movement. Redrawn and modified from Arber and Costa (2018).

precision actions involving independent finger movements in primates, like playing the piano or knitting, may depend on direct corticospinal projections to motoneurons. Transection of the corticospinal axons, leaving the rest of the cortical control of the basal ganglia, midbrain, and brainstem intact, leads to a loss of independent finger movements, but the other part of the motor repertoire remains largely intact (Porter, 1987; Lawrence & Kuypers, 1968a, 1968b).

1.4 THE BLUEPRINT OF THE VERTEBRATE MOTOR SYSTEM IS 500 MILLION YEARS OLD

The first vertebrate group that diverged from the evolutionary line leading to mammals are the cyclostomes with two sister groups, the lamprey and the hagfish. What is common between the detailed organization of the lamprey nervous system of today and the mammalian nervous system most likely existed at the point when the lamprey diverged from the mammalian line, around 560 million years ago, at the time of the Cambrian explosion (Kumar & Hedges, 1998). The lamprey nervous system has been subject to a detailed analysis on all levels, from the spinal cord to the forebrain, during the last few decades. It shows remarkable similarities to the mammalian lineage, as we will discuss next, and has recently been reviewed in some detail (Suryanarayana et al., 2021a, 2022; Grillner, 2021). The gross organization with a spinal cord, brainstem, midbrain, and forebrain had been established early, as well as the presence of cranial motor nuclei and dorsal root ganglions and ventral roots in the spinal cord. For instance, Sigmund Freud (1878) as a very young neurologist established the sensory organization of the lamprey spinal cord before directing his attention to the complexity of the human subconscious and the presumed importance of dreams. In this section, I will compare the lamprey nervous system with that of mammals, particularly the parts that are involved in the control of motion.

1.4.1 The Lamprey Cortex: Pallium

The lamprey pallium/cortex had been considered as mainly an olfactory structure without layers since the early 1900s. This turned out to be wrong, as we started to explore the role of the forebrain in the context of how movements are selected by the brain. We found that the small structure corresponding to the mammalian neocortex, called the "pallium," was a three-layered cortex with a molecular layer and two cellular layers, and with the same proportion of inhibitory GABAergic interneurons and excitatory glutamatergic interneurons as in mammals (figure 1.9a–e). Moreover, stimulation of the lamprey pallium could elicit eye, orienting, and locomotor movements in a circumscribed motor area in which projection neurons targeted downstream motor centers in the tectum, midbrain, brainstem, and spinal cord in the same way as in mammals

(Ocana et al., 2015; Suryanarayana et al., 2017, 2020), and with the same type of monosynaptic connections and synaptic transmission (figures 1.9f,g). The properties of the projection neurons resemble the mammalian counterparts regarding cellular properties and their dendrites extending into the molecular layer, as is characteristic for the mammalian pyramidal neurons.

We could subsequently show that there is a visual area in the lamprey pallium/cortex with a retinotopic representation (different parts of the visual field are represented by separate neurons located in different parts of the cortex). Between the motor area and the visual area is a somatosensory area with different parts of the body represented in different locations (Suryanarayana et al., 2020; figure 1.9h). Thus, the lamprey cortex (dorsal pallium) has the same general organization with a motor area, a visual area, and a somatosensory area in sequence. The thalamic relay neurons are specific to the sensory input from the retina and somatosensory input, respectively, and terminate on neurons corresponding to the mammalian layer 4 neurons in the cortex. The sensory neurons of the olfactory mucosa target the mitral and tufted-like cells in the olfactory bulb (figure 1.9i), and project in turn and separately to different parts of the olfactory cortex (ventral pallium), as in mammals (Suryanarayana et al., 2021b).

In conclusion, the overall organization with the motor and sensory representation in the pallium/cortex is present in the lamprey, but in three instead of six layers as in the neocortex. The olfactory cortex has three layers in the lamprey and in its mammalian counterpart. The cell types and proportion of γ-aminobutyric acid (GABA) interneurons and glutamate neurons are similar. This means that the lamprey cortex provides a blueprint of that of mammals, with the basic building blocks of the cortex in place, but with the number of neurons being orders of magnitude smaller and thereby having less computational power. Whether there is a hippocampal area is not yet clear, but a neural solution for spatial navigation is most likely required.

1.4.2   The Basal Ganglia of Importance for Selection of Behavior: Conserved from Lampreys to Mammals

When we started to investigate the lamprey basal ganglia, we anticipated a simplified structure that could aid in the selection of behavior through

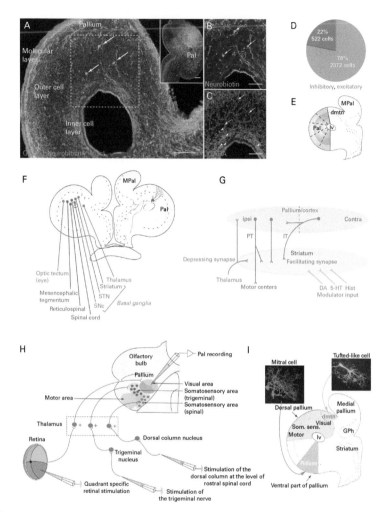

Figure 1.9

Cell types and organization in the lamprey pallium/cortex. *A.* Retrogradely labeled IT type cells (arrows) following neurobiotin injection in the contralateral pallium (inset). *B, C.* The IT-type cells (b, arrows), from the region of the dotted square in *A* do not express GABA (*C*, arrowheads). Note the three layers—the molecular layer and the outer and inner cellular layers. *D.* Overall percentage and number of GABAergic and non-GABAergic pallial neurons. *E.* Schematic of the lamprey pallium with segments, indicated with dotted lines, used for cell counting (Pal, pallium; MPal, medial pallium; lv, lateral ventricle). *F.* Schematic of a transverse section of the lamprey pallium indicating the efferent targets of pallial projection neurons (PT-type, pyramidal tract type). *G.* Projections of PT- and IT-type pallial neurons in the lamprey. *H.* Summarizing schematic of the lamprey dorsal pallium, showing retinotopic visual areas, somatosensory areas, and motor areas, as well as the retinal, trigeminal, and dorsal column nucleus afferents relayed via distinct subpopulations of thalamic neurons. *I.* The olfactory system in lamprey resembles that of mammals in many respects with a dual efferent system from the main olfactory bulb conveyed via tufted-like and mitral cells. The latter target the olfactory cortex located in the ventral pallium, whereas the former target a separate limited region, the dorsomedial telencephalic nucleus (dmtn). Note the different morphologies of the mitral (magenta) and tufted-like (blue) cells. mya, million years ago. From Suryanarayana et al. (2022).

an action on downstream motor center. We were very surprised to find an organization that in considerable detail resembled that of mammals.

The general organization of the basal ganglia in figure 1.10a applies to both lampreys and mammals. The input level, the striatum, the output level SNr/GPi, the intrinsic nucleus globus pallidus externa (GPe) and the subthalamic nucleus (STN) are the major building blocks of the vertebrate basal ganglia. The afferent projections to the striatum are from the same two types of neurons in the pallium/cortex, pyramidal tract (PT) projection neurons, and intratelencephalic (IT) neurons, as well as from the thalamus. The striatum consists mostly of projection neurons subdivided in those expressing dopamine receptors of the D1 or D2 type, which are the origin of those that subserve initiation of behavior (D1), the direct pathway, and those that instead counteract movements (D2) and form the indirect pathway (Stephenson-Jones et al., 2011, 2012a; Grillner & Robertson, 2016). The dopamine innervation originates from the lamprey SNc and is activated by salient stimuli that originate in part from the conserved lamprey SC, the tectum, as in mammals (Pérez-Fernández et al., 2017), as discussed further in chapter 4. At least two of the subtypes of the mammalian interneurons in the striatum exist in the lamprey. Furthermore, the connectivity, the transmitters employed, and the neuropeptides expressed (e.g., substance P and enkephalin; figure 1.10b) are conserved. The related circuitry of the lateral habenula and the circuitry involved in the control of the level of activity in the dopamine neurons are virtually identical (Stephenson-Jones 2012b, 2013, 2016; figures 1.10c,d).

In conclusion, the basal ganglia, a major part of the circuitry involved in the control and selection of behavior, is conserved regarding the general organization with subnuclei, transmitters and intrinsic connectivity and cellular membrane properties (see also chapter 4). The fact that the basal ganglia have remained practically unchanged is presumably related to the fact that the design has functioned well and there has been no evolutionary pressure to modify the circuitry—or rather any change/mutations induced have been for the worse. However, the number of neurons has increased many times over in mammals.

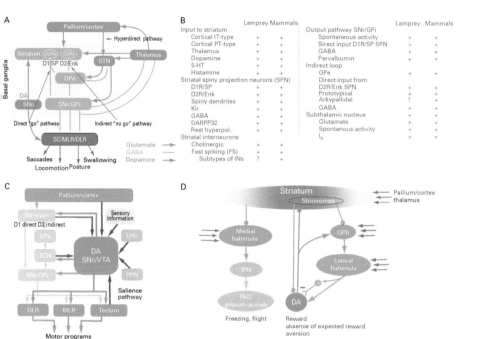

Figure 1.10

The lamprey basal ganglia. *A.* The organization of the basal ganglia. The striatum consists of GABAergic neurons, as do the GPe, GPi, and SNr. The SNr and GPi represent the output level of the basal ganglia, and they project via various subpopulations of neurons to the tectum/SC, the MLR, and diencephalic locomotor region (DLR), and other brainstem motor centers, as well as back to thalamus with efference copies of information sent to the brainstem. The direct striatal projection neurons (dSPNs) that target SNr/GPi express the dopamine D1 receptor (D1) and substance P (SP), while the indirect striatal projection neurons (iSPNs) express the dopamine D2 receptor (D2) and enkephalin (Enk). Also indicated is the dopamine input from the SNc (green) to striatum and brainstem centers. Excitatory glutamatergic neurons are shown in pink and GABAergic structures in blue. *B.* A table showing the key features of the basal ganglia organization that are found in mammals and the lamprey. *C.* The SNc connectome in the lamprey and mammals. The efferent and afferent connectivities of the SNc are virtually identical. *D.* The striatum targets both the medial and lateral habenulas. The medial habenula sends projections to the interpeduncular nucleus (IPN) and further to the PAG/griseum centrale. In mammals, the PAG triggers a variety of fixed action patterns related to freezing and flight. The lateral habenula is engaged in the control of the various modulator systems and receives input from the glutamatergic part of globus pallidus (GPh), which in turn receives inhibition from the striosomal compartment of the striatum. The striosomes, GPh, and lateral habenula itself receive input from the pallium, thalamus, and other structures. From Suryanarayana et al. (2022).

1.4.3 The Thalamus and Hypothalamus

The thalamus serves in all vertebrates as a sensory relay nucleus that forwards information from the retinal cells to neurons in the visual areas in the cortex/ pallium, and correspondingly, somatosensory information is channeled to the somatosensory areas. The thalamus also mediates preprocessed information through the higher-order thalamic nuclei, such as major input from the tectum/SC that is forwarded to the cortical/pallidal areas involved in visuomotor coordination. The thalamus receives input from the output nuclei of the basal ganglia, which in turn target a specific thalamo-striatal projection that provides feedback to the striatum. These components exist in both the lamprey and mammals (Suryanarayana et al., 2020, 2022).

In mammals, the hypothalamus is involved in the control of the endocrine system through the hypophysis, the diurnal rhythm from the suprachiasmatic nucleus, defensive behaviors (freezing, escape, aggression), reproductive behaviors (sexual and maternal behaviors), and ingestive behaviors (food and fluid intake, foraging). These are all essential aspects for survival, and they are considered innate (see chapter 2). In the lamprey, the organization of this aspect of behavior has not been studied in any detail. But the control of the hypophysis is present, as well as the general organization in subnuclei (Nieuwenhuys & Nicholson, 1998), the expression of several peptides (e.g., galanin), and a conserved histamine projection (Brodin et al., 1990). The PAG receives major input from the hypothalamus to its various subcompartments that conveys many aspects of the defensive, ingestive, and reproductive behaviors in mammals. The lamprey counterpart of PAG, the *griseum centrale*, also receives input from the hypothalamus (Olson et al., 2017).

In conclusion, the thalamus in the lamprey has in principle the same design as in mammals, with a relay function to cortical sensory areas, and thalamostriatal projections and input from the tectum/SC are important for visuomotor coordination. The details of the hypothalamic control of behavior in the lamprey remain to be studied, but the overall organization, based mostly on anatomy, seems to resemble that of mammals.

1.4.4 The Midbrain Roof, the Tectum, Is Concerned with Events in the Immediate Surrounding Space

The roof of the midbrain is called the "tectum" in most vertebrates, but in mammals, it is called the "superior colliculus (SC)." The tectum/SC is concerned with what happens in the immediate surrounding space. Events activating different parts of the visual space will activate selective parts of the retina, which in turn will send axons to the superficial part of the tectum/SC, arranged such that a retinotopic map of the surrounding space is formed. If a salient stimulus activates a specific spot in the retina, it will then activate a specific area in the tectal map. The tectal neurons in this area can then activate tectal output neurons, which in turn elicit a saccadic eye movement or a head-orienting movement to direct the gaze toward the salient object that gave rise to the gaze shift. Subsequently, it will be determined which action to take. Moreover, the selected population of tectal neurons activated will strongly inhibit the neurons in the surrounding tectal area, which inhibits other potential movements. This arrangement of the tectum is present in all vertebrates that depend on vision, whether in lampreys, eagles, or primates (Isa et al., 2021).

Information about the surrounding space is also provided by sounds. When humans or other animals are in a pitch-dark environment, auditory information provides an auditory spatial map that is fed into the tectum at a level just deeper than and spatially aligned with the visual map. This arrangement also exists in the lamprey, but electrosensation provides the map of the surrounding space, mediated by receptors on the head and body, which is aligned with the visual map. If visual and electrosensory stimuli originating from the same point in space occur at the same time, they facilitate each other. Concurrent stimuli originating from different parts of the surrounding space instead inhibit each other in the lamprey, and correspondingly in mammals with visual and auditory stimuli (Kardamakis et al., 2016; Stein & Stanford, 2008). The processed information in the tectum/SC is also forwarded to the thalamus and further to the higher-order visual areas in the cortex/pallium from the lamprey to primates (Isa et al., 2021; for details, see chapter 3).

In conclusion, the design of the circuits in the roof of the midbrain that respond to events in the immediate surrounding space remain similar throughout vertebrate phylogeny, although vision can be combined with other senses such as hearing, electrosensation, and even infrared radiation in different vertebrate groups. However, the retinal map is further refined in mammals, with retinal afferents providing a number of types of visual information separated into different layers of the visual map.

1.4.5 Cerebellum

The cerebellum is present in all vertebrate groups, from fish to mammals, but in the lamprey, there is only a tiny bridge of neurons across the rostral part of the fourth ventricle that has been considered to be a precursor of the cerebellum (Sugahara et al., 2021b). It is, however, uncertain, whether the presumed cerebellar circuitry is functional. A recent single-cell RNA sequencing study indicates that the lamprey is devoid of cerebellar cell types (Lamanna et al., 2022). The cerebellum is well developed in elasmobranchs (sharks and rays), the next group to be formed in the phylogenetic tree.

1.4.6 Cranial Nerves and the Senses

Olfaction and vision are well developed in the lamprey, as well as their central connectivity as mentioned previously. This applies also to the sensory input from the trigeminal area. The vestibular apparatus has only two semicircular canals, but they are arranged so that they can sense rapid movements in three planes. Receptors signaling the head position are located in a vestibular sac, resembling the situation in mammals (sacculus and utriculus).

On the motor side, the innervation of the eye muscles through the oculomotor, trochlear, and abducens nerves are similar between the lamprey and mammals with some limited differences (Fritzsch et al., 1990). The trigeminal innervation controls the oral muscles, and there is also a facial nucleus (the seventh cranial nerve, NVII). The glossopharyngeal and vagus nerves carry the efferents to the respiratory muscles and have sensory gustatory components.

The organization and roles of the different cranial nerves were established before the lamprey line of evolution diverged from that leading to mammals.

1.4.7 Brainstem

The lamprey brainstem has a group of descending reticulospinal neurons subdivided into an anterior, middle, and posterior nucleus, which convey monosynaptic excitation to spinal motoneurons and interneurons, as in the pontine reticulospinal pathway in mammals (Grillner & Lund, 1968; Rovainen, 1974; Ohta & Grillner, 1989). They are active during the initiation of locomotion, postural adjustments, and steering (see also chapter 2). There is also a crossed and uncrossed vestibular projection to the rostral spinal cord and a 5-HT projection to the spinal cord, like the raphe-spinal pathway in both mammals and lamprey (Brodin et al., 1986). In the lamprey, there is a very limited supply of noradrenergic innervation and no coeruleo-spinal projection. These are components that are represented in mammals, where the number of nuclei and control systems has been markedly expanded (e.g., chewing occurs only in mammals).

Of course, the addition of forelimbs and hindlimbs in tetrapods during evolution requires further control via discrete pathways, as does the changing type of locomotion from body undulations to tetrapod walk and gallop. The versatility of the limbs to be used for reaching and grasping and the manipulation of food and other items evolve gradually until it finally reaches perfection in the control of the human hand.

The brainstem respiratory centers that drive the rhythmic respiratory movements are to a large degree conserved. In contrast, the respiratory movements used to extract oxygen from water in the gill pouches of the lamprey differ markedly from those used when breathing air through lungs, as in tetrapods (see chapter 2; Cinelli et al., 2013; Gariépy et al., 2012).

In conclusion, the lamprey brainstem contains much of the basic vertebrate design as in tetrapods, with vestibular control mechanisms, the reticulospinal contribution, and a 5-HT system. There is also the intrinsic organization with the processing related to the various cranial sensory and motor nuclei and pattern generator networks. The noradrenergic system has not evolved, and neither the refined control of the nonexisting appendages.

1.4.8 Spinal Cord

The lamprey spinal cord has approximately 100 segments and has been reviewed in some detail by Robertson et al. (2021). The dorsal roots convey sensory information from the segmental dorsal ganglia, and in addition, there are two types of large intraspinal neurons that convey touch and pressure sensation, respectively. One branch of their axons extends through the dorsal roots to innervate the skin. Their central branch is divided into a descending and an ascending branch, as in mammals. The latter joins the dorsal column, which terminates in the dorsal column nucleus, which in turn forms the medial lemniscal pathway that forwards information to the thalamus (Suryanarayana et al., 2020, 2022). The axons of the motoneurons exit through the ventral roots to target the segmental muscle fibers. The ventral funiculus contains the large axons of the reticulospinal neurons that allow fast corrections, while the lateral funiculus contain thinner axons from the lateral part of the brainstem and spinoreticular projections. The lamprey nervous system is unmyelinated, and therefore some axons have a large diameter that increases the speed of transmission along the axon.

Vertebrates from fish to mammals have an LMC that controls the muscles of the forelimbs and hindlimbs, respectively, and an MMC that controls the neck and trunk muscles (Dasen & Jessell, 2009; see figure 1.11). The lamprey, with no appendages, has only an MMC. The motor neurons are subdivided into those that control the dorsal and ventral parts of the myotome and the slow muscle fibers separately (Teräväinen & Rovainen, 1971). They have somewhat different extensions of the dendritic arbor that reach from the cell body to the larger part of the lateral and ventral funiculus, but not into the dorsal columns. Their membrane properties are typical for motoneurons in general. The spinal cord contains the neuronal networks underlying locomotion, composed of excitatory glutamatergic interneurons with ipsilateral projections to motoneurons and crossed inhibitory commissural interneurons (which will be dealt with further in chapter 2). As in other vertebrates, the central canal is lined with liquor-contacting cells that (at least in the lamprey) sense any deviation of pH from neutral; and when activated, they release somatostatin (SST), causing an inhibition of the motor circuits that reduces motor activity and thereby contributes to the restoration of a neutral pH (Jalalvand et al., 2016, 2022). It seems likely that these homeostatic pH-controlling mechanisms of cells around the central canal also exist in other vertebrates, but that remains to be shown.

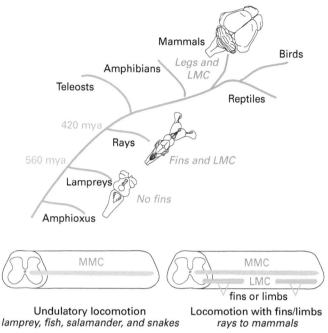

Figure 1.11

Vertebrate evolution from amphioxus to mammals, with the time indicated when the different groups became separate from the line of evolution leading up to mammals. The paired appendages emerged in elasmobranchs together with the LMC. At the bottom are the LMC, important for locomotion with paired appendages, and the MMC, critical for the control of the entire body and undulatory locomotion. LMC, lateral motor column; MMC, medial motor column; mya, million years ago. From Grillner and El Manira (2020).

In conclusion, the lamprey spinal cord is designed in a similar way to that of other vertebrates, except that it lacks an LMC and has no appendages to control. The LMC, with motoneurons and interneurons, is well developed in the next phylogenetic group to appear some 420 million years ago, the elasmobranchs/skates (Jung et al., 2018). The lamprey nervous system is unmyelinated, but axons can have a thin membranous sheath around the axonal membrane.

1.4.9 The Lamprey in an Evolutionary Perspective: Before and After

From the forebrain to the spinal cord, all essential features exist at the lamprey stage of vertebrate evolution, and thus they are assumed to have existed when the lamprey line of evolution became separate from that leading to mammals more than 500 million years ago (Kumar & Hedges, 1998). Of course, the number of neurons is small in all parts of the lamprey nervous system compared to most other vertebrates, and therefore the behavioral repertoire is much more limited. The data summarized here and in greater detail in other sources (e.g., Suryanarayana et al., 2021a, 2022; Grillner & Robertson, 2016), show that beyond any reasonable doubt, the outline, the blueprint of the vertebrate nervous system, had evolved early in vertebrate phylogeny. This finding is also supported by molecular/genetic studies of transcription factors and precursors expressed in the lamprey brain (Sugahara et al., 2021a; Lamanna et al., 2022), which further demonstrate the conserved structure of the lamprey nervous system.

The difference between the lamprey brain and that of the preceding phylogenetic stages, the tunicates and the amphioxus, a cephalochordate, is dramatic. The amphioxus has a spinal cord, a vestigial brainstem, and a possible telencephalic domain, the pars anterodorsalis, with glutamatergic, GABAergic, and dopaminergic neurons. It remains to be ascertained if this presumptive telencephalic domain, in the dorsal locus of the hypothalamus alar plate, is equivalent to the corresponding vertebrate prosomere (Lacalli, 2022; Puelles & Rubinstein, 2015). There is, however, no evidence of major forebrain structures, such as the thalamus and the midbrain structures in vertebrates. Between the comparatively simple organization of the amphioxus nervous system and the lamprey with a blueprint of the vertebrate nervous system, a radical, complex evolution must have taken place. We may never learn the details of this astounding development since they remain hidden and possibly are absent in fossil records.

2 EXECUTION OF MOVEMENT: A PALETTE OF CPGS AND MOTOR CENTERS FROM MIDBRAIN TO SPINAL CORD

2.1 INTRODUCTION

From a general description of the motor repertoire of vertebrates and the motor networks that exist in different parts of the brainstem and spinal cord provided in chapter 1 and collectively referred to as the "motor infrastructure" (figure 1.6 in chapter 1; also see Grillner, 2003), I will now move to the next analytical level by discussing the intrinsic function of these networks. How does each of them do its job? They achieve this by combining a battery of synaptic, cellular, and molecular building blocks and forming a network, as will be discussed next.

To understand a neuronal network is more complex than one would think at first sight. One needs first to define the different groups of neurons that are essential for generating the motor pattern concerned, and then discover how they talk to each other. Excitation is mostly mediated by glutamate synapses that act via a subset of α-amino-3-hydroxy-5-methyl-4-isoxazolepropionic acid (AMPA) and N-methyl-D-aspartate (NMDA) receptors that can induce a variety of postsynaptic effects and lead to different forms of short-term plasticity. Inhibition is mediated mostly by glycine in the networks of the spinal cord, although γ-aminobutyric acid (GABA) is also used in the dorsal horn. In the brainstem, it is a mix between GABA and glycine, and in the forebrain, it is exclusively GABA. It is critical to define the connectivity and synaptic interaction among the neurons to reach a firm understanding. If one finally has obtained the wiring diagram of the network and the properties of the synapses, one is off to a good start but still far from having sufficient information.

The membrane properties are also very important to consider since neurons have very different properties. Some may have pacemaker or plateau properties; others can be interneurons that faithfully relay the information from their input structures. Some may respond with pronounced spike frequency adaptation, and still others may display spontaneous activity. The properties that each cell has depends on which specific combination of ion channels is expressed—namely, subtypes of Na^+, K^+, Ca^{2+} and calcium- and sodium-dependent K^+ channels (see Grillner, 2003). Furthermore, the shape of the soma-dendritic tree contributes to the processing that takes place at different parts of the dendritic tree, such as plateau potential or synaptic plasticity. When one has so many different factors that covary, detailed simulation is a very useful complementary tool in conjunction with a precise experimental analysis (see the discussion that follows). These microcircuits subserving many different functions are sometimes referred to as "motor primitives" (see the definitions of these terms in section 1.3.1 of chapter 1) and are to a large degree evolutionarily conserved (i.e., present during vertebrate phylogeny).

In this chapter, we will discuss a number of these microcircuits, how they become activated, their modes of operation, and what is known about their intrinsic functions. I earlier made the analogy of an orchestra represented by the various microcircuits, while the forebrain corresponds to the conductor, which controls the activity of each member of the orchestra.

2.2 CPG NETWORKS PRODUCING LOCOMOTOR, RESPIRATORY, AND CHEWING MOVEMENTS AND RELATED BEHAVIORS

In all vertebrates, networks producing the rhythmic motor activity underlying locomotion and respiration exist in the spinal cord and in the brainstem. These networks contain the information required to activate the muscles taking part in the step cycle or the respiratory motion, and they are activated in an appropriate sequence. These networks are often referred to as "central pattern generator (CPG) networks." Most vertebrate groups swallow the food directly; only mammals have evolved a chewing CPG network, located in the brainstem.

2.2.1 Cellular and Synaptic Mechanisms for Burst Generation

What are the cellular and synaptic mechanisms that are used to generate a rhythmic burst activity such as in the CPGs discussed here (Grillner et al., 2005)? Consider a population of excitatory neurons that display recurring burst activity (figure 2.1). Essentially, the depolarizing phase can be supported by synaptic glutamatergic interactions between the excitatory neurons, combined with mechanisms that tend to promote depolarizing plateaus.

Figure 2.1a shows an example of how the various cellular mechanisms can support rhythmic activity. With an excitatory background drive from neurons that initiate burst activity, the neurons become depolarized, and the voltage-dependent properties of NMDA channels or persistent Na$^+$-channels will boost the depolarization that can be maintained during a plateau.

The plateaus would depend on voltage-dependent NMDA channels and persistent Na$^+$-channels. Why does the depolarization-phase during the burst not just continue? This can be accounted for by a gradual accumulation

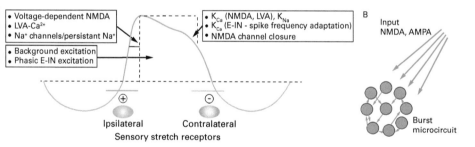

Figure 2.1

A. Several factors contribute to the initiation, maintenance, and termination of the depolarizing phase. In addition to conventional synaptic, voltage-dependent NMDA receptors, low-voltage-activated Ca^{2+} channels (LVA-Ca^{2+}) and Na$^+$ channels might be activated. Ca^{2+} enters the cell through these channels, activates K$_{Ca}$ channels, and initiates a progressive hyperpolarization leading to closure of the NMDA channels. The initiation of depolarization is facilitated by activation of ipsilateral excitatory stretch receptors, whereas its termination is partially a result of activation of contralateral inhibitory stretch receptors. E-IN, excitatory interneuron. Dashed line indicates the resting membrane potential. *B.* Schematic representation of a burst-generating circuit.

of Ca^{2+} and Na^+ intracellularly, which leads to a progressive activation of calcium- and sodium-dependent potassium channels (K_{Ca} and K_{Na}), which will repolarize the cell membrane. As the membrane becomes hyperpolarized, the NMDA and persistent Na^+-channels will close, and the neurons will become silenced for a period until the hyperpolarization declines, and a new burst will start due to the background excitatory drive. The postburst hyperpolarization can also lead to an activation of hyperpolarization-activated depolarizing currents, such as I_h and low-voltage activated Ca^{2+}-currents, which will induce a postinhibitory rebound that will promote the depolarization and the initiation of a new burst. The synaptic excitatory interaction among the neurons in the population ascertains that they remain synchronized. The molecular and synaptic mechanisms mentioned here form a toolbox that can be utilized in a variety of circuits within the central nervous system.

2.2.2 The Respiratory CPG Contributing to Breathing and Related Orofacial Behavior

Respiration is critical for all creatures, and always served by a CPG network (figure 2.2). In mammals, breathing is characterized by an active inspiration, followed by an expiration that may be passive at low rates of breathing (recoil of the rib cage), but at higher rates, there is an actively controlled expiration.

The core of the respiratory CPG is the pre-Bötzinger complex (pBC), a group of neurons located ventrally in the brainstem at the level of the XII cranial nerve nucleus. If the pBC is inactivated, respiration stops (Del Negro et al., 2018; Ashhad & Feldman, 2020). The pBC consists of a mixture of inhibitory glycinergic and subgroups of excitatory glutamatergic neurons. One of the glutamatergic subgroups coexpresses somatostatin (SST). The network can produce rhythmic burst activity when inhibition is blocked, and thus it can operate without inhibition. The SST-negative excitatory neurons form a network that, through mutual excitation within the population, is thought to generate the rhythm, and after a burst, the neurons become refractory, start to be active again due to their inherent excitability, and generate a burst through mutual excitation. They excite the SST-positive neurons in the pBC, which in turn activate downstream neurons that activate the inspiratory motoneurons supplying the diaphragm and the intercostal

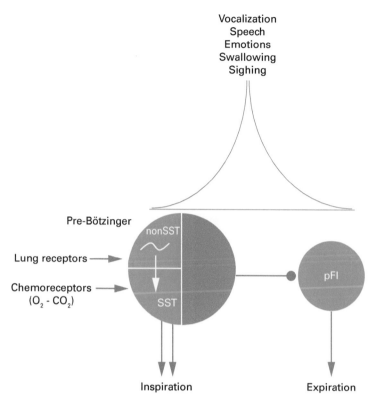

Vocalization
Speech
Emotions
Swallowing
Sighing

Pre-Bötzinger

nonSST

Lung receptors →

Chemoreceptors →
$(O_2 - CO_2)$

SST

pFl

Inspiration

Expiration

Figure 2.2
Organization of the respiratory network generating inspiratory and expiratory movements. The pre-Bötzinger complex contains one population of cells that generates the burst pattern that does not express somatostatin (nonSST), a set of output neurons expressing somatostatin (SST), and a set of inhibitory neurons. Expiration is generated by the lateral parafacial nucleus (pFl). Red represents glutamatergic neurons and blue represents inhibitory neurons.

muscles. The inhibitory neurons are thus not needed for burst generation to occur, but they nevertheless take an active part and slow the burst generation.

How is the expiratory phase generated? The lateral parafacial (pFl) nucleus, located rostral and ventral to the pBC, is responsible, and when active, it drives the expiratory motoneurons. The pBC provides inhibition of the pFl during each burst, ascertaining that an alternation with the pFl will occur (figure 2.2). The pFl will not be active at low rates of breathing, but it starts to contribute as the metabolic demand increases, resulting in an active expiration (Del Negro et al., 2018).

Recent data (Anderson et al., 2016) suggest that there is a third group of neurons that are active at the interface between the inspiratory and expiratory phase, constituting the postinhibitory complex (PiCo). This group of neurons are located close to the pBC, but in a somewhat more dorsal and rostral location. It is argued that processes such as swallowing, vocalization, and speech, which need to halt the breathing temporarily, act via the PiCo. The PiCo was not included in figure 2.2 since the connectivity is as yet unclear.

The respiratory CPG ticks on at a rate decided by the metabolic demands as signaled by chemoreceptors in a brainstem nucleus and in the periphery, but it is subject to a variety of influences, and the CPG is sensitive to all inputs. For the cycle-to-cycle regulation, vagal afferents signaling the expansion of the lungs during each inspiration will act on the CPG and help terminate the inspiration at the appropriate time (Clark & von Euler, 1972).

Several times an hour, humans, as well as rodents and other mammals, sigh without noticing. The sigh consists of a deep and long-lasting inspiration, the purpose of which is to expand the lungs and fill alveoli that may have collapsed. This is controlled by a small set of 200 neurons in the nearby retrotrapezoid nucleus that release either of two neuropeptides, neuromedin B or gastrin-releasing peptide (Li et al., 2016). They activate peptidergic receptors on cells within the pBC and induce a sigh. A knockout of the receptors abolishes sighing without affecting other aspects of breathing. Sighing occurs at very low rate, and the mechanism by which the activity of the sigh-inducing peptidergic neurons are controlled is not yet known.

In anticipation of the metabolic needs, breathing is also increased directly when a locomotor command is turned on to initiate locomotion, which is before the actual needs for more oxygen has been manifested both in mammals and the lamprey (DiMarco et al., 1983; Gariépy et al., 2012). Emotional factors also affect the rate of breathing, and it can become shallow or combined with a deep breath as part of an emotional state. Furthermore, one can hold one's breath for some time voluntarily until the metabolic drive takes over, and also voluntarily increase the rate of breathing. During vocalization or speech, the airflow is controlled actively by an action on the respiratory CPG and by the vocal cords in the larynx.

The respiratory CPG in the lamprey has a very similar design to that of the pBC regarding the excitatory interaction with both AMPA and NMDA receptors, and a cholinergic and peptidergic contribution. Essential parts of the rhythm-generating network were thus present early in vertebrate phylogeny and appears to have been evolutionarily conserved (Cinelli et al., 2013; Missaghi et al., 2016).

In conclusion, the respiratory CPG ascertains that the metabolic demands are met, but it is sensitive to a variety of behavioral demands that can instantaneously modify each breath, both in duration and amplitude as for instance during speech.

2.2.2.1 Relation between respiration and chewing, licking, swallowing, and other patterns of orofacial behavior

The respiratory pBC interact with and is affected by several neural circuits that control licking, chewing and vibrissae movements, as summarized in figure 2.3. In rats, the basic rate of breathing is around 2 Hz, while rhythmic licking occurs at 5–7 Hz and continues while breathing, but the frequency of licking is modulated

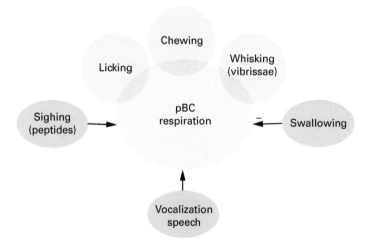

Figure 2.3
The pre-Bötzinger complex (pBC) interacts with three related rhythmic CPGs (light blue), and is also inhibited by the circuits generating swallowing, sighing, and vocalization, including speech.

somewhat in each respiratory cycle. The licking involves rhythmic opening of the jaw and an alternating protrusion and retraction of the tongue, and it is controlled by a CPG located in the intermediate reticular formation dorsomedial to the pBC (Wiesenfeld et al., 1977; Travers et al., 1997). The tongue can also be directed to the left or right, as in a reaching movement, and rodents can be trained to respond in a direction-specific manner with the tongue (see chapter 5).

The related CPG for chewing operates at around 4 Hz, when active, and it is located in the pons, near the trigeminal motor nucleus, with part of the circuitry in the intermediate reticular formation and adjacent parvocellular reticular formation just dorsal to the licking CPG area (Moore et al., 2014). Neurons in these areas include premotor interneurons to the jaw closure and opener motoneurons, respectively, and can display rhythmic bursting when activated (Dellow & Lund, 1971; Kolta et al., 2007; Moore et al., 2014). Chewing is characterized by alternation between the jaw opening (digastric muscle) and closing (masseter muscle) and involves tongue movements, which will redistribute the food in the mouth while chewing. However, there are many variations in chewing movements—one can chew mostly on the left or the right side or in a grinding fashion. This has led to a proposal of a CPG composed of a palette of local interacting CPGs (figure 2.4) like that of the mammalian locomotor CPG. The detailed connectivity of the CPG is only partially understood, but the rhythmic activity of neurons of the trigeminal sensorimotor circuit is reported to depend on persistent Na^+-channels that operate at low extracellular Ca^{2+}.

In addition to the neurons in the chewing CPG network, astrocytes appear to play an important role in pattern generation. When activated by glutamate, the local astrocytes can release a calcium chelator, S100beta, which lowers the extracellular levels of Ca^{2+}, which promotes the activation of persistent Na^+-channels, and thereby the rhythmic activity in the network (Morquette et al., 2015; Condamine et al., 2018; Slaoui Hasnaoui et al., 2020). The contribution of this mechanism is shown by the fact that rhythmicity is blocked when the action of S100beta is blocked by antibodies specific to S100beta. This provides an important demonstration of a dynamic interaction between astrocytes and neurons in the control of chewing network function. Since the normal function of astrocytes is to regulate

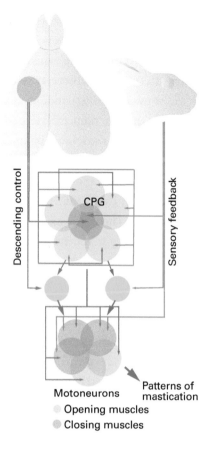

Descending control

CPG

Sensory feedback

Motoneurons

Patterns of mastication

○ Opening muscles
○ Closing muscles

Figure 2.4

A conceptual scheme of the masticatory control system. Included is the variability observed in jaw movements, which can emphasize activity on one side of the mouth or the other. The CPG is driven from the brainstem and cortical areas and is thought to have a number of modules, which allow different emphases on the activation of different muscle groups, as discussed in this chapter in the context of the locomotor CPG. There is similarly a sensory control indicated, which can affect both the CPG level and the motoneurons. Courtesy of James Lund.

the local extracellular space, such as for glutamate, any interference with astrocyte function will likely lead to changes in network function. However, the present demonstration of the effect of S100beta is more specific.

When the food has been chewed, it is brought to the posterior part of the oral cavity and swallowing is initiated. It starts by closure of the respiratory tract and inhibition of respiratory and chewing movements, and then the swallowing CPG is activated (Jean, 1990, 2001). This CPG, located along the nucleus tractus solitarius, produces a sequence of muscle contractions that will bring the food bolus down to the esophagus and then forward it to the ventricle through the gastric sphincter, a process that takes several seconds. To allow the food to pass downward, the distal motoneurons are inhibited, while more proximal muscles in sequence propel the food forward. The CPG can operate without sensory input, but the sensory input can halt the swallowing process, as when a fish bone has been ingested. The process of swallowing can be reversed as in vomiting, with inhibition of breathing and activation of the abdominal muscles (Jean, 2001; Lang et al., 1993).

Yet another respiration-associated behavior comprises the rhythmic movements of the whiskers, particularly in rodents. The whisking movements are controlled from a CPG located dorsomedial to the pBC in the ventral part of the intermediate reticular formation (Kleinfeld et al., 2014; Moore et al., 2013, 2014), mediated via the facial nerve to the muscles moving the vibrissae. The whisking CPG can operate independently but input from the pBC often makes the two rhythms become synchronous (see figure 2.3). The whisking behavior has become an important experimental model since the input from the individual vibrissae are represented in separate compartments within the "barrel cortex" and provides discrete input to the striatum and other parts of the cortex. The whisking provides spatial information regarding the immediate surroundings processed at many levels of the nervous system from the cortex to the midbrain, which is particularly important in animals that move around in dark environments, like many rodents.

In conclusion, the brainstem contains a set of orofacial CPGs that control various aspects of feeding, such as chewing, licking, and swallowing, which all interact in one way or another with the respiratory pBC. The whisking CPG also belongs to this category. The networks are mostly located in close association with each other in the lower brainstem and form a center for the control of orofacial movements.

2.2.3 The Brainstem–Spinal Cord Control of Locomotion

In section 1.2 of chapter 1, I discussed the many versions of locomotion that occur among vertebrates, from fish to primates. What is common among them is that in all cases, a CPG network extending within the spinal cord generates the motor pattern underlying locomotion in a specific species. Moreover, the CPG does not work in isolation; it receives continuous information from a variety of sensors that adapt the movements to external events as the animal moves through a complex terrain or swims in water currents (Grillner & El Manira, 2020).

Lamprey/fish CPGs involve the entire spinal cord and can generate forward undulatory locomotor movements that can easily be reversed, as in backward swimming (Grillner et al., 1976). For tetrapods, each limb and the trunk are coordinated, the movements of each limb can be controlled independently, and the coordination between them can be varied from walk and trot to gallop. Even animals with a spinal transection can perform walking and galloping movements on a treadmill directly after a spinal transection if noradrenergic agonists are administered, or in the chronic spinal state (Forssberg & Grillner, 1973; Andersson et al., 1981; Rossignol et al., 2006).

2.2.3.1 Supraspinal command centers for locomotion

In all the vertebrates investigated, locomotion can be elicited by stimulation of an area referred to as the "mesencephalic locomotor region (MLR)" (Shik et al., 1966; Ryczko et al., 2016), located in the transition between the mesencephalon and the pons. This area contains several neuronal structures, and it has been unclear for a long time which of them are the most important. Recent studies in the mouse have shown that neurons in the cuneiform nucleus make important contributions, and the effects are mediated to the spinal cord CPGs via neurons in the lateral paragigantocellular nucleus (LPGi) (Kiehn, 2016; Roseberry et al., 2016; Capelli et al., 2017; Caggiano et al., 2018; Ferreira-Pinto et al., 2021). Direct spinally projecting neurons from the MLR do not elicit locomotion when activated selectively, but instead they cause body lengthening and rearing (Ferreira-Pinto et al., 2021). However, with concurrent activation of locomotion, this may be part of an integrated behavior leading to stiffening of the body. The pedunculopontine nucleus (PPN) is also located within the area of the MLR and its cholinergic neurons may contribute to locomotor control, particularly exploratory movements. The MLR thus forms an important integrated center for the control of both locomotion and rearing. In the lamprey, the MLR similarly activates locomotion through bilateral effects on reticulospinal neurons (Ryczko et al., 2016). There is also an area within the diencephalon that elicits locomotion, called the diencephalic locomotor region (DLR), which contains neurons projecting to the lower brainstem, corresponding most likely to the zona incerta in mammals and the ventral thalamic area in the lamprey (Orlovsky, 1969; El Manira et al., 1997). These two areas, MLR and DLR, can thus be used by different forebrain structures to initiate locomotion in many contexts, such as escape or foraging.

2.2.3.2 The intrinsic function of locomotor CPGs

It has long been known that the CPG in mammals can produce the full motor pattern underlying the four phases of the step cycle (Grillner & Zangger, 1975). The next-level question was to find out how the spinal CPG actually produces the motor pattern. To understand a microcircuit, one needs to know the following:

- The identity of the neurons, which are part of the CPG
- Their membrane properties
- The synaptic interaction—transmitters and receptor subtypes
- How the CPG is activated

This seems straightforward, but in reality, it is a very demanding task. Therefore, we decided in the late 1970s to use the lamprey for the experiments, instead of a mammal. The lamprey belongs to the oldest group of now-living vertebrates, and the lamprey spinal cord is more accessible, has much fewer neurons, and can be maintained in vitro, while the locomotor CPG can be turned on experimentally.

2.2.3.3 The lamprey locomotor CPG

The lamprey spinal cord turned out to be a good choice. It is thin (only 200–300 μm in depth), a natural slice, and could be used in vitro, and therefore methodologically advantageous (figure 2.5). Equally important, the locomotor pattern could be induced by electrical stimulation of the MLR or pharmacologically. We could then study the locomotor CPG in action under in vitro conditions, and after some years, conclude that populations

A Dorsal view of the spinal cord

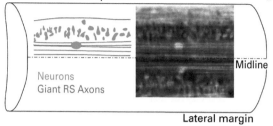

Neurons
Giant RS Axons

Midline

Lateral margin

B The suction mouth

C Swimming Adult Lamprey *(Lampetra fluviatilis)*

Figure 2.5
A. The lamprey has a thin (200–300µm), flattened spinal cord that is well suited for imaging and electrophysiology, and the larger neurons are visible under the microscope. *B.* The lamprey is a jawless vertebrate with a sucker mouth. *C.* The lamprey swims with an undulatory wave propagated along the body.

of excitatory premotor interneurons (EINs) in each segment represent the core of the burst-generating circuit. The EINs in turn activate ipsilateral motoneurons and glycinergic commissural interneurons (CINs) that inhibit contralateral interneurons and motoneurons (figure 2.6b; see also Buchanan & Grillner, 1987). The locomotor burst activity was initiated by the locomotor command acting via the excitatory reticulospinal neurons from the different brainstem nuclei that activated the CPG by targeting both EINs and CINs through activation of both NMDA and AMPA receptors (Ohta & Grillner, 1989; figure 2.6c). The burst activity will be facilitated by the activation of the voltage-dependent NMDA receptors that contribute to the depolarizing plateau (Wallén & Grillner, 1987; Alford & Williams, 1989). The burst termination depends on the cellular properties of the CPG neurons and is accounted for by a gradual intracellular accumulation of Ca^{2+} and Na^+, leading to a progressive activation of calcium- and K_{Na}-channels, which will hyperpolarize the EINs (figure 2.5d; El Manira et al., 1994; Wallén et al.,

2007) and inactivate other voltage-dependent processes. The burst will then be followed by a hyperpolarizing phase when they become refractory, and subsequently due to the depolarizing drive from the locomotor command, they again become depolarized, and the voltage-dependent NMDA channels open and contribute to the initiation of the next burst. The CPG can generate burst activity even when the glycinergic inhibition is inactivated, as is the case for the respiratory CPG (as discussed previously) and the mammalian locomotor CPG (Cangiano & Grillner, 2005; Grillner & El Manira, 2020). Normally, however, the commissural interneurons are responsible for the reciprocal inhibition that results in alternating activity between the left and the right sides.

When the cells within the CPG and their synaptic interactions had been characterized, it was still difficult to know if the experimental findings could account for the network operation. One way to test if the experimental findings can indeed produce the locomotor pattern is to use simulations based on detailed information of the neuronal and synaptic properties established experimentally. In this case, we simulated the segmental CPG in a realistic way to test if the model network can reproduce the biological function. Figures 2.6a and 2.6c show an example, with a model of the EINs at the cellular level with appropriate morphology, expression of ion channels, and afterhyperpolarization, and each model neuron is reproducing in detail their biological counterpart (current, frequency relation etc). A population of model EINs (figure 2.6b) were then connected with each other and driven by excitation from the brainstem mediated via NMDA and AMPA receptors (figure 2.6c,d). The intrinsic cellular mechanisms that contribute to burst termination are illustrated in figure 2.6e. One could then show that the model network could reproduce the activity of the biological network and generate burst activity in the same broad frequency range (0.2–10 Hz) as the biological network (figure 2.6g).

What we have discussed so far is only the fast synaptic interaction via ionotropic glutamate, glycine, and GABA receptors. In addition, modulatory metabotrobic receptors activated by 5-HT, GABA ($GABA_B$), and glutamate (metabotropic glutamate receptors [mGluRs]), and peptides can fine-tune CPG activity by acting on molecular components of the cells

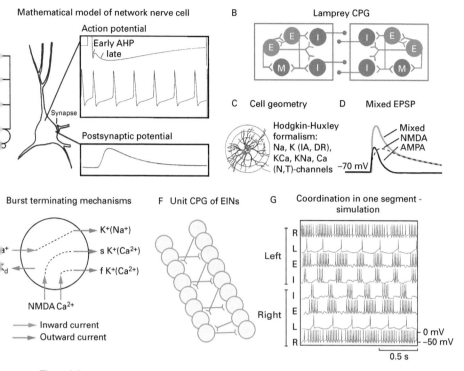

Figure 2.6

The lamprey segmental network—cellular properties explored in simulations of the lamprey spinal locomotor networks. *A*, *C*. The morphology of the CPG neurons is captured using a five-compartment model consisting of an initial segment, a soma, and a dendritic tree. Active ion currents are modeled using a Hodgkin-Huxley formalism. Both ion channels involved in spiking behavior (Na$^+$, K$^+$) and slower Ca^{2+} or potassium-dependent processes (K$_{Ca}$, K$_{Na}$) are modeled based on available data. Spike frequency can be regulated by the AHP. *B*. Shows the connectivity within the spinal CPG (red glutamatergic neurons, blue glycinergic inhibitory neurons, and green motoneurons). *D*. Fast synaptic transmission is included in the form of excitatory glutamatergic (AMPA and voltage-dependent NMDA) and inhibitory glycinergic inputs. *E*. Main ionic membrane and synaptic currents considered to be important during activation within the CPG network. Slower processes can cause spike frequency adaptation, such as Ca^{2+} accumulation during ongoing spiking and resulting activation of K$_{Ca}$ (see also the discussion in this chapter). *F*. Unilateral CPG activity can be evoked in local networks of EINs. This basic EIN network can sustain the rhythm seen in ventral roots in vitro during evoked locomotor activity. The left–right alternating activity requires the presence of contralateral inhibition provided by glycinergic interneurons (as discussed in this chapter). *G*. Simulation of the segmental network in operation. Abbreviations: E/EINs, excitatory interneurons; EPSP, excitatory postsynaptic potential; I, inhibitory interneurons; M, motoneurons.

within the CPG, such as ion channels (Grillner, 2003). For instance, the presence of 5-HT markedly facilitates regular rhythmic activity in the locomotor CPGs, extending from the lamprey to mammals. Although this represents an important aspect of CPG function and could be the subject of a book of its own, we will not discuss this important aspect of neuromodulation further in this context (Grillner, 2003).

2.2.3.4 Generation of the undulatory wave: Intersegmental coordination in the lamprey

So far, we have considered the burst-generating capacity on the segmental level, whereas in reality, the lamprey spinal cord has 100 segments, all of which are coordinated so that a wave from head to tail pushes the animal forward during swimming, with a fixed phase lag between each of the segments (figure 2.7). When the lamprey or fish swims, the speed is increased mainly by increasing the frequency of swim cycles, in the case of the lamprey from around 0.2 to 10 Hz, and in the case of a trout, up to 25 Hz from head to tail(!). The phase lag between each of the segments in the lamprey is around 1 percent of the cycle duration, regardless of whether it is very short or long, with the result that the phase lag along the body will remain the same at low or high speed. The speed of the traveling wave along the body, however, will

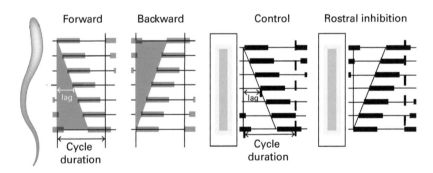

Figure 2.7
Intersegmental coordination during swimming in lamprey. An undulatory wave is transmitted along the body from head to tail, while it is reversed during backward swimming. There is a corresponding delay in the activation of the muscle segments, as well as the direction of the wave. In the isolated spinal cord, the head-to-tail wave can be generated, and the wave can be reversed by adding excitation to the most caudal segments or (conversely) inhibiting the most rostral segments.

increase inversely to the duration of the swim cycle. The underlying neural network depends on the same components as in the segmental CPG, with the addition that the inhibitory commissural neurons have long, intersegmental, descending axons that make synapses over perhaps 20 segments, while the EINs have briefer intersegmental axonal branches (not more than 5–6 segments), as indicated in figure 2.7.

Figure 2.8 shows that a network, incorporating the intersegmental network from head to tail consisting of 10,000 neurons (100/segment), can generate activity with the appropriate phase lag along the spinal cord (Kozlov et al., 2009, 2014). Equally important, if the excitability of the first few segments were increased, the phase lag along the entire spinal cord could increase to even more than 2 percent, corresponding to two undulatory waves over the entire body. Conversely, if the same segments were inhibited to some degree, so that they would have a lower excitability than the remaining part of the spinal cord, the wave reversed in direction over the entire spinal cord, resulting in a tail-to-head direction, corresponding to backward swimming. By changing the excitability of a few rostral segments gradually, the phase lag can be graded from 2 percent to −2 percent. The normal range would be from 1 percent to −1 percent, but during rapid changes of speed, the phase lag can vary over some cycles.

One important take-home message is that by tinkering with the excitability of just a few segments of a large network, one can affect the operation of the entire network. Another important conclusion was that the variability observed in each population was an important design feature. When we designed the model network, we also included the variability in cellular properties that had been observed experimentally. For example, somewhat smaller interneurons would have higher input resistance and be recruited first when depolarized, and conversely large neurons would join somewhat later during each burst. When variability existed, the network operated in a very stable manner over a large frequency range, whereas if all neurons in each population had identical properties (an average neuron), the network produced a less stable rhythm in a narrow frequency range. It is intuitively easy to realize that a gradual recruitment and derecruitment of neurons during each burst will result in a more stable rhythm. Thus, the variability in

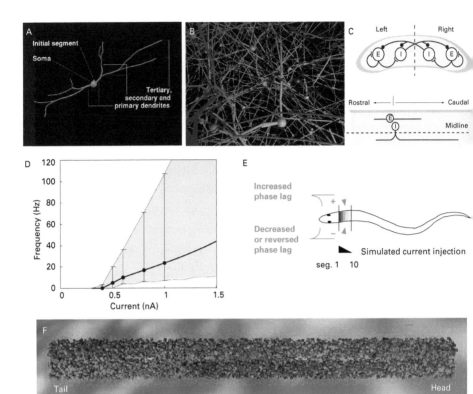

Figure 2.8

Full-scale simulation of the spinal locomotor network of lamprey with 10,000 neurons in the spinal locomotor network. *A.* The morphology of neurons was similar to their biological counterparts, and they were simulated with the variability observed experimentally. *B.* The density of neurons within part of the neuronal pool. *C.* Organization of the synaptic connections in transverse and longitudinal views of the spinal cord. Caudally directed projections dominate. Excitatory (E) neurons project ipsilaterally (i.e., their axons do not cross the midline), and inhibitory (I) neurons project contralaterally. *D.* Mean firing rate (thick line) as a function of the somatic current injection. The shaded area shows maximum variation of the spike frequency in the simulated population. *E.* The magnitude of the intersegmental phase lag varies continuously with changes of the excitability in a few rostral segments. A lowering of excitability reverses the phase lag to backward swimming in all segments, while a neutral stage leads to the normal forward phase lag, and an enhanced excitability increases the phase lag to values greater than normal. *F.* The overall activity in the simulated spinal cord during locomotion. Each dot signifies the activity of one neuron. The color code in this image is blue for hyperpolarized, red for firing neurons, and yellow for depolarized but not yet firing neurons. Note that when one side is active, the other is inhibited, and the overall activity corresponds to the phase lag observed in vivo.

cellular properties of a class of neurons is not a mistake of nature, but rather a design feature.

2.2.3.5 Simulation of actual swimming and steering

So far, we have considered the activity at the network level, but the network in turn controls a body that interacts with the surrounding world (in this case, the viscoelastic properties of water). To include these properties in the simulations, we performed a simulation of a lamprey-shaped body with segmental muscles with appropriate viscoelastic properties controlled from the network by the segmental output from motoneurons (Ekeberg & Grillner, 1999; Kozlov et al., 2009, 2014). Moreover, the water environment has been included. Figure 2.9 shows the outline of the lamprey and the rostrocaudal excursions of its swimming movement, very close to the natural movements of the lamprey (Williams et al., 1989).

These simulations were very instructive in bridging the gap between the neural network activity and actual behavior, such as swimming as the speed of

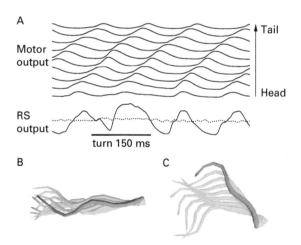

Figure 2.9

Electrical and mechanical dynamics in the model. *A.* Phasic and tonic components of reticulospinal (RS) and motor outputs from the right side. An output signal of neuron population is the firing frequency computed by leaky integrators with time constant equal to the cell membrane time constant. Timing of the turn command is shown by the horizontal bar. *B, C.* Shape of the body during forward swimming and a lateral turn caused by activation of the turn command in tectum. Modified from Kozlov et al. (2014).

movement increased (Grillner et al., 2007; Kotaleski et al., 1999). Moreover, we simulated steering left and right by adding excitatory drive to the network on either side by incorporating the crossed and uncrossed outputs from the tectum to reticulospinal neurons, which could make the motoneurons on one side somewhat more active and induce a turning movement. The more excitation that was added, the deeper the resulting turning movement would become (Kozlov et al., 2014). With the increased activity on one side, there was a corresponding attenuation of the activity on the contralateral side.

2.2.3.6 Propriospinal control of the locomotor CPG by stretch receptor neurons: Biology and simulation

Stretch-receptor neurons located along the margin of the spinal cord sense the lateral undulations during swimming since the spinal cord is displaced in each swim cycle. These neurons (also called "edge cells") have fine ramifications in the lateral margin that sense mechanical distortion. They are of two types: one is glutamatergic and excites ipsilateral CPG neurons, and the other is glycinergic and inhibits contralateral neurons (Grillner et al., 1984; Di Prisco et al., 1990). If passive, locomotor-like movements are imposed on the isolated spinal cord during fictive locomotion, the locomotor activity becomes entrained at frequencies that can be both somewhat lower or higher than the original rest rate when no movements were imposed (Grillner et al., 1981).

As described previously, when the swimming movements were simulated, the introduction of the edge cells in the control scheme of the local CPG does not affect locomotor movements. However, if the locomotor movements were challenged (e.g., with unpredicted water currents and higher water speeds), the feedback became important. A simulated lamprey without edge cells cannot handle the perturbations and cannot move forward as planned, whereas with the sensory feedback, it can indeed compensate and move forward (see figure 2.10).

2.2.3.7 The segmental locomotor CPG in zebrafish

Later work in the zebrafish has shown that EINs can be subdivided into three groups (subgroups of the genetically defined V2a interneurons, as shown in figure 2.11) (Song et al., 2018, 2020). One group activates slow motoneurons and is responsible for slow swimming. Another activates intermediate

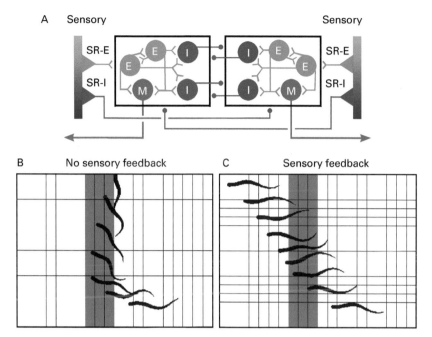

Figure 2.10

The segmental CPG network is indicated together with stretch-receptor neurons (i.e., edge cells) that sense the locomotor movement. *A.* One excitatory stretch receptor subtype acting on the ipsilateral side of the CPG and a contralateral that inhibits contralateral CPG neurons. *B.* The swimming lamprey model is coordinating its swimming perfectly well without sensory feedback, but when it meets an area in which the water is rapidly moving toward the lamprey, it is unable to cope with the perturbation (gray area). *C.* The sensory feedback (stretch receptors) is included in the network, and it is then able to swim through the perturbed area. Abbreviations: E, excitatory interneurons; I, inhibitory interneurons; M, motoneurons, SR-E, excitatory stretch receptor neurons; SR-I, inhibitory stretch receptor neurons. Modified from Ekeberg and Grillner (1999).

motoneurons/muscle fibers, and a third activates the very fast motoneurons that innervate muscle fibers required during escape swimming. The slow EINs have membrane properties conducive to generating rhythmic activity, and they mutually excite each other and weakly excite the intermediate type of EINs. They also excite each other, as well as the slow EINs, thereby ascertaining synchronous burst activity in all active EINs.

The intermediate type of EINs is recruited, with a stronger descending drive from the brainstem, and fast EINs are involved only transiently during escape. The zebrafish spinal locomotor CPG is currently the best-characterized

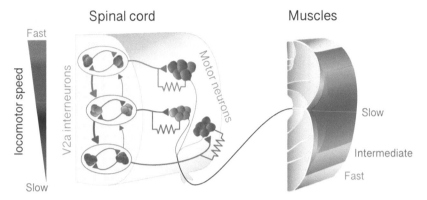

Spinal cord　　　　　　　　Muscles

Fast

locomotor speed

V2a interneurons

Motor neurons

Slow

Slow

Intermediate

Fast

Figure 2.11

The modular zebrafish CPG. As described in this chapter, the excitatory V2a interneurons are further subdivided into three modules, each controlling fast (blue), intermediate (green), and slow (red) motoneurons and muscle fibers. Each module consists of excitatory interneurons, some of which also have gap junctions in their synapse on motoneurons (as indicated by the resistor in the drawing). The gap junctions play an important role in providing feedback between motoneurons and interneurons, thereby amplifying the chemical synaptic transmission when depolarized. To the left, it indicates that at slow speeds, the slow (red) module is recruited, and with increasing speed the intermediate (green) and the fast (blue), respectively. Courtesy of Abdel El Manira.

vertebrate CPG. The sensory control of the segmental CPG is from receptors that sense the lateral movements of the body located at the margin of the spinal cord and connected to intervertebral ligaments (Picton et al., 2021). An inactivation of these sensory propriospinal neurons affects the locomotor movements, which become slower, and their amplitude tends to become larger, showing that the proprioceptive control is an integral part of the locomotor network. The basic design of these circuits in the zebrafish resembles those of the lamprey, and their stretch receptor neurons that interact directly with the locomotor CPG.

2.2.3.8　The mammalian locomotor CPG

The mammalian limb CPG, with its four phases, has been more difficult to crack. The neonatal mouse preparation that generates mostly flexor-extensor alternation, however, has allowed an analysis of some interneurons and how they interact (Goulding, 2009; Kiehn, 2016). The categories of interneurons in the spinal cord can be genetically defined in subgroups from dorsal to ventral, as shown in figure 2.12a (Jessell, 2000). The alternation between flexors and extensors is produced by the genetically defined V1 and V2b inhibitory

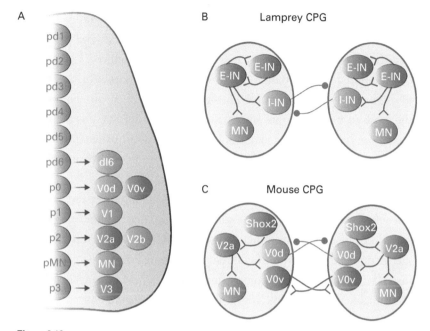

Figure 2.12

Schematic representation of the precursor groups in the developing vertebrate spinal cord and the lamprey and mouse CPG. *A.* The spinal cord is subdivided into six dorsal and five ventral progenitor domains (pds), giving rise to different interneuron populations and motoneurons. To the right are indicated the classes of neurons referred to in the text in the context of locomotion. The color code is blue for inhibitory, red for excitatory interneurons, and green for motoneurons. *B.* The lamprey CPG with the same color code. *C.* The mouse CPG. Courtesy of Abdel El Manira.

interneurons. If both types are inactivated, the alternation disappears and flexors and extensors become synchronous, while the bursting continues. The bursting process itself does thus not require inhibition (Rancic & Gosgnach. 2021; Grillner & Kozlov, 2021).

There are three types of excitatory interneurons: V2a interneurons, a non-V2a population called Shox2 subpopulation, and HB9 (Homeobox). The Shox2 interneurons are rhythmically active in phase with either flexors or extensors. They are electrically coupled, may contribute to generating the rhythmic burst activity, and are thought to excite the V2a interneurons, which in turn activates the motoneurons (figure 2.12c; Dougherty et al., 2013; Ha & Dougherty, 2018). Inactivation of selected types of interneurons has not yet resulted in conclusive answers.

Thus, we do not yet fully understand the process of generating the burst activity in the mammalian locomotor CPG, nor the next-level problem of how the four-phase locomotor pattern of the limb is produced. One factor to consider is that each population of interneurons needs to be further divided into subgroups that supply individual groups of motoneurons. For instance, the inhibitory neurons that supply the ankle extensors are not the same as those supplying the knee extensors. There is thus a further fractionation within each subpopulation that needs to be considered, which makes the analyses more demanding.

The coordination between the left and right hindlimbs and forelimbs can shift between walking, trotting, or galloping. A very interesting finding regarding the coordination between the limbs is that two sets of crossed interneurons, the V0d and V0v, are engaged (figure 2.12c), one during walking and the other during trotting. Knocking out either of them will make the mice unable to trot or walk, respectively (Talpalar et al., 2013).

2.2.3.9 The unit burst generator concept for the locomotor CPG: A versatile organization

The locomotor CPG needs to be versatile and be able to contribute to walking forward and backward and adapt a whole variety of modifications of the locomotor pattern. The alternative of having, instead, one CPG for each possible variation of locomotion seems unattractive and costly (e.g., in terms of the number of neurons needed).

The CPGs for each of the four limbs can be combined in different ways to account for the different gaits (figure 2.13a). The next-level question is to consider whether the limb CPG could be further subdivided into unit CPGs that could be recombined to generate the types of coordination required, such as backward, forward, or sideways locomotion (figure 2.13b).

An important experimental finding was that the ability to generate rhythmic activity is distributed locally over the entire spinal cord, as shown with optogenetic activation of glutamatergic interneurons. Unilateral bursting can be produced in one segment, and one can even elicit bursting exclusively in either flexor or extensor motoneurons (Hägglund et al., 2013). These findings agreed with earlier, less detailed findings in the cat by Grillner

(1981, 1985), which showed that bursting could be distributed in different parts of the spinal cord, and furthermore that either flexors or extensors could be active in isolation. These studies therefore suggested a conceptual model of a flexible CPG organization, in which each group of synergists at a given joint were controlled by a unit CPG. The diagram in figure 2.13b shows the concept, with one burst-generating unit CPG for the hip extensors (HE) and another for the hip flexors (HF) and the same for the knee, ankle, and foot. At each joint, the unit CPGs for flexors and extensors are reciprocally connected, while the connectivity between the joints could be varied. For forward locomotion, the extensor unit CPGs work together, while for backward locomotion, hip and knee extensors inhibit each other, as the phase relation between the hip and lower limb will need to change between forward and backward locomotion. This organization can also generate the actual motor pattern, with the knee flexors being active before other flexors, as they are responsible for the liftoff at the end of the support phase by letting the ankle flexor (AF) and hip flexor (HF) unit CPGs inhibit the knee flexor (KF) unit CPG. In that way, the KFs become active before the other flexors. The brief toe extensor, the extensor digitorum brevis (EDB), is also a muscle that will have a burst at touchdown before the other extensors.

To further explore this network configuration, we simulated the network (Grillner & Kozlov, 2021) in an unorthodox way. We used as unit CPGs the segmental populations of EINs (see figure 2.5b, showing the lamprey CPG) that are known to have the ability to generate burst activity in isolation, and then connected them as indicated in the scheme. The simulations in figure 2.13c show that the burst pattern resulted in a pattern similar to that observed experimentally, with alternation between flexors and extensors in the hip and ankle, but with the knee flexors active at the transition from extensor to flexor activity (i.e., liftoff phase) and in the other transition, and similarly with EDB, which is also active particularly when the foot is placed on the ground. This shows that the pattern of activity matches that of the normal locomotor pattern, at least for the first approximation.

As the speed of locomotion increases, the support phase is shortened markedly, while the flexion phase remains rather constant in animals as well as humans (Goslow et al., 1973; Grillner et al., 1981). We simulated this

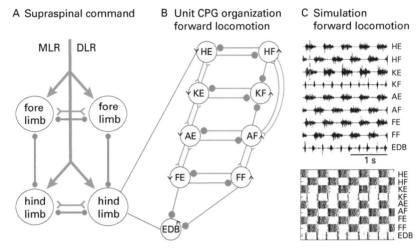

Figure 2.13

Systems of interacting unit CPGs—intralimb coordination—forward locomotion. *A.* The diagram to the left shows the four limb CPGs of a tetrapod and possible modes of coordination (in phase or alternation for the forelimbs and hindlimbs, respectively). The CPGs are turned on by the descending drive from the MLR or the DLR. *B.* Within each limb CPG, there is most likely a further subdivision in unit CPGs controlling the synergists at one joint, such as hip (H), knee (K), ankle (A), and foot (F) extensors (E) or flexors (F). The extensor digitorum brevis (EDB) has a particular pattern. The normal pattern of activity results from the interaction between the unit CPGs at various joints. The advantage is that the unit CPGs may be recombined as in backward or forward walking, in the same way as with the limb CPGs that can be recombined in the various gaits. Circles indicate inhibition, and forks/triangles excitation. *C.* Exploratory simulation of locomotor activity, with a network arranged as in the diagram in which each unit CPG is designed in a similar way to the lamprey unit CPGs consisting of 100 excitatory interacting neurons and the interaction between the 9-unit CPGs arranged as in *B.* The output of the model network captures essential features of the locomotor output. Modified from Grillner and Kozlov (2021).

condition with the same network by gradually increasing the excitatory drive to the extensor-unit CPGs, while letting the drive to the flexor-unit CPGs remain similar. As can be seen in figure 2.14a, the simulation produced a constant flexion phase, while the duration of the extension phase shortened dramatically. Thus, this network configuration can reproduce this important aspect of the adaptation to speed present in mammals and other tetrapods.

What about versatility? Can the network be made to generate the backward-walking motor pattern? If the connectivity is changed so the mutual excitation between hip and knee extensor in forward locomotion is

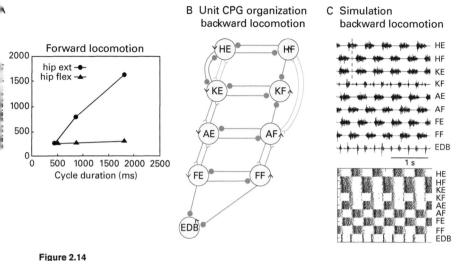

Figure 2.14

Systems of interacting unit CPGs—intralimb coordination—in backward locomotion. *A.* The graph shows that the flexion phase can be kept constant at different cycle durations, while the extension phase can vary markedly when the excitatory drive to the extensor-unit CPGs is varied, although the drive to the flexor-unit CPGs is kept constant. *B.* Note that the connectivity pattern between the hip-unit CPGs and the lower-limb-unit CPGs has been modified. Weakened synaptic projections are shown with thin lines. *C.* Same representation of the motor pattern to the muscles as in figure 2.11 during simulated backward locomotion. The image below indicates the activity in individual interneurons. Abbreviations: AE, ankle extensor; AF, ankle flexor; EDB, extensor digitorum brevis; FE, foot extensor; FF, foot flexor; HE, hip extensor; HF, hip flexor; KE, knee extensor; KF, knee flexor. Modified from Grillner and Kozlov (2021).

replaced by reciprocal inhibition and the excitation between hip and ankle flexors is also removed (figures 2.14b and 2.14c), alternation between the hip and knee extensors is produced, which would correspond to the coordination required for backward locomotion.

The unit CPG organization of the locomotor network would thus account for the flexibility of the motor pattern and allows a subtle control compared to just forward and backward locomotion. It allows for utilizing part of the unit CPG organization for other movements such as wiggling the big toe or performing isolated ankle or knee movements, as I wrote in Grillner (1985). In planning for movements around individual joints, one can argue that the simplest solution would be to use the relevant part of the infrastructure available in the locomotor CPG organization. This most likely applies to the evolutionarily conserved rhythmic scratch reflexes performed by the hindlimbs, as well as the forelimb self-grooming movements (Grillner & El Manira, 2020). Rather than designing a new microcircuit at the spinal level, when producing the spinal aspect of reaching, grasping, or hand-to-mouth coordination, it would seem likely that parts of the available spinal machinery is reused. As discussed next, the different groups of neurons in the brainstem, such as the lateral rostral medulla (latRM) are important to consider in this context, but the processing at the spinal level is only partially known. This view is reinforced by the observations of Hägglund et al. (2013), which reported that fractions of the locomotor pattern can be induced by local activation of glutamatergic interneurons along the spinal cord.

2.2.3.10 Escape reactions

It is critical for all vertebrates to react to impending threats by escaping (see also section 2.4). In most mammals, escape will result from an activation of one compartment of the periaqueductal gray (PAG) and downstream MLR, and in lower vertebrates most likely through the *griseum centrale,* the analogous structure to PAG. Escape reactions in fish consist of an initial C-shaped bend of the body to change direction, followed rapidly by escape swimming at maximal frequency. It can be elicited through the Mauthner cells, which project monosynaptically to contralateral motoneurons (Faber et al., 1989). Recent findings in zebrafish show that Mauthner cells also activate a specific subset of cholinergic V2a interneurons that selectively target the fast subtype of motoneurons by gap junctions located on the axon hillock, thereby bypassing the dendrites and soma of motoneurons, resulting in a minimal delay (Guan et al., 2021). These interneurons amplify the escape reaction markedly, and their ablation eliminates the escape command reaction.

2.3   A BRAINSTEM CENTER FOR COORDINATION OF REACHING AND GRASPING MOVEMENTS IN THE LATERAL RETICULAR MEDULLA

To be able to reach toward and grasp an object, mostly with the forelimbs, is an important part of the tetrapod motor repertoire. Thus, a mouse can reach toward a pellet and then grasp it and bring it to the mouth. This pattern of behavior is served by innate circuitry located in the brainstem and spinal cord. As so often is the case, it has been long assumed that reaching depended exclusively on projections from the motor cortex to the spinal cord.

Recent experiments from the Arber laboratory (Esposito et al., 2014; Ruder et al., 2021), however, have shown that in the mouse, a group of neurons in the lateral rostral medulla (latRM) serves as a center for reaching and grasping coordination (figure 2.15). One set of neurons in the latRM projects directly to

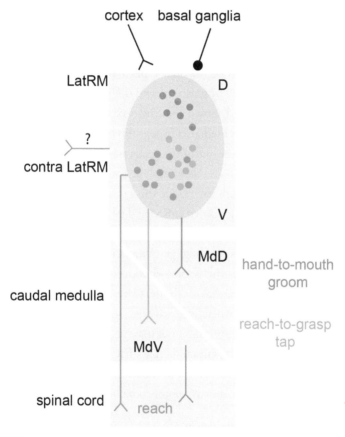

Figure 2.15
Center in the latRM coordinating reaching grasping coordination. It contains one population of cells that elicit reaching, another that via the MdV population controls the hand movements in grasping, and a third (MdD) that elicits hand-to-mouth movements. LatRM, lateral rostral medulla; MdD, medullary reticular formation dorsal part; MdV, medullary reticular formation ventral part.

the spinal cord, and their selective activation leads to a reaching movement that is not combined with other types of movements. Some of these neurons have a bias toward the direction of the reaching movement (e.g., lateral versus medial).

Another set of latRM neurons projects to the ventral part of the nearby brainstem medullary reticular formation ventral part (MdV). This nucleus projects directly to motoneurons and interneurons in the spinal cord that enable the paw to open as the first part of a grasping movement. Activation of MdV elicits part of a grasping movement, and an activation of the two together presumably produces a reach-to-grasp movement. Finally, a third group of neurons in the latRM elicits a hand-to-mouth movement acting through neurons in the dorsal part of the brainstem—namely, medullary reticular information (MdD). The three types of neurons are activated in succession: reaching, followed by grasping, and then retracting (to the mouth). Older work in the cat had shown that the execution of the reaching command at the spinal level is mediated by a group of propriospinal neurons in segments C3–C4, while the grasping signals are mediated at the segmental level (Alstermark et al., 2007; Alstermark & Isa 2012; Pivetta et al., 2014) and several descending pathways excite this group of C3–C4 neurons.

Taken together, this means that latRM and related groups of neurons form a command center in the brainstem–spinal cord for eliciting all aspects of this important behavior that is reaching, grasping, and bringing objects such as a piece of food to the mouth. This is somewhat analogous to the locomotor command centers in the mesencephalon. The next-level question is which parts of the nervous system determine when the reach-grasp infrastructure should be called into action. The recent demonstration of massive projections from the output nucleus of the basal ganglia, substantia nigra pars reticulata (SNr) (McElvain et al., 2021), to this part of the reticular formation may suggest that the basal ganglia is involved. Projections from the motor cortex are also important to consider.

2.3.1 Cortical Neurons Involved in Reaching

It has long been known that in primates and cats, neurons in the motor cortex are activated during reaching (Georgopoulos et al., 1986; Georgopoulos, 1986; Yakovenko & Drew, 2015). In the primate experiment, the monkey was sitting in a chair with the hand pointing to a central position, and when

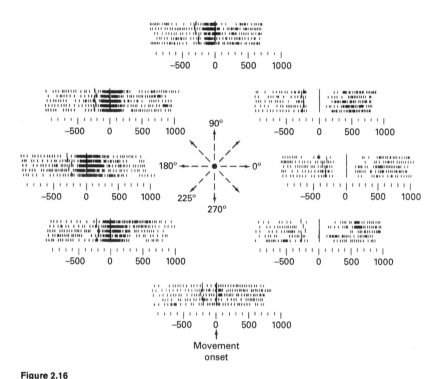

Figure 2.16

Individual cortical neurons are broadly tuned to the detection of movement. Raster plots show the firing pattern of a single neuron of movements in eight directions. A monkey was trained to move a handle to eight locations, represented by light-emitting diodes, arranged radially in one plane around a central starting position. Each row of tics in each raster plot represents activity in a single trial. The rows are aligned at zero time (the onset of movement). The center diagram shows the directions of the eight movements. The cell fires at relatively high rates during movements made in directions ranging from 90 to 250 degrees. From Georgopoulos et al. (1982).

any one of eight lights arranged in different directions around the center were turned on, the monkey should reach out toward the light. Neurons in the arm area of the motor cortex were recorded during each reaching movement. These neurons that were active during reaching were broadly tuned, but for each of the eight directions, the population vector of the neurons pointed in the appropriate direction (see figure 2.16). The neurons were also activated, but to a lesser degree, with movements to nearby directions. When reaching movements in the opposite direction were performed, they were instead inhibited. By recording from populations of neurons, the direction of the reaching movement performed could be independently predicted. The cortical neurons are thus activated when the monkey performs reaching movements. Moreover, they are in a broad sense specific for the direction of the reaching movement. Many of the neurons in the motor cortex project to the basal ganglia (striatum), brainstem, and some do so to the spinal cord. It seems likely that the latRM command center for reaching/grasping is also present in monkeys (figure 2.15). If so, one would assume that the reaching neurons in the motor cortex impinge on the monkey latRM and could thereby elicit a reaching-grasping movement sequence. Another likely input to the latRM would be from the output nuclei of the basal ganglia, such as the SNr (McElvain et al., 2021), which project to this general region. This level of control will be further discussed in chapters 4 and 5, considering the role of the forebrain in controlling movement.

A further sophistication takes place when handling food, which requires interactions between the grasping paw and oral processing. When a mouse has grasped a piece of food, it can further handle the object by an interaction between the forelimb paw and the mouth, like removing the envelope of a seed before eating it. In another example, a mouse fed a piece of spaghetti will use both paws to orient it so it can be ingested effectively (An et al., 2022; Mohan et al., 2019). This complex process involves pyramidal neurons in two related orofacial motor areas in the frontal lobe, and it would seem likely that they act through or interact with the neurons in the latRM.

As noted in chapter 1, in reaching/grasping movements, the rodent and human strategy for performing the movements is very similar, which has led to the conclusion that the underlying neural circuits would be conserved (Sacrey et al., 2009). Since this behavior is present in all tetrapods, a processing center like that of the rodent latRM may be present in vertebrates ranging from frogs to primates.

2.3.2 Timing of Reaching Movements Requires the Ability to Predict the Location of a Moving Target

There are, however, many demanding types of processing underlying the timing of a reaching movement toward a moving object, so that the paw/hand reaches the object at the appropriate time. To achieve this, one cannot aim the reaching toward the actual location of a moving object at time zero, but rather predict the location of the object when the reaching movement has been completed. Consider, for instance, when responding to a serve in tennis, one needs to make an unconscious prediction of the trajectory of the ball and position the arm extended with the racket in an appropriate position. This ability to make predictions is critical for all animals when (for instance) hunting for prey. Even fish make such timing predictions when attacking prey. A pike waits in one place for small fish to pass by, and then it aims at the future location of the fish at the precise moment when it can attack the victim (Kashin et al., 1977).

Human neonates reach out with their arms toward colored objects passing close by, and at week 36, they are close to the reaching quality and precision of the adult. They then can make a prediction of where the object should be located when they can capture it (von Hofsten & Lindhagen, 1979). This ability to predict the location of a moving object is clearly an important part of the behavioral repertoire of vertebrates, from fish to humans, and critical for many types of movement control.

2.4 THE PAG CHANNELS COMMANDS FROM THE HYPOTHALAMUS AND AMYGDALA

2.4.1 The PAG: A Command Region for Vocalizations, Escape, Freezing, Lordosis, Maternal Behavior, and Pain Relief

In mammals, the area around the aqueduct that extends between the fourth and the third ventricle is referred to as the "periaqueductal gray (PAG)" (Vieira et al., 2011; Faull et al., 2019; Subramanian et al., 2021). In birds, fish, and lampreys, it is called the *griseum centrale* (Wullimann et al., 1996; Dubbeldam & den Boer-Visser, 2002; Olson et al., 2017). This is a comparatively large area, and in humans, it extends over 25 mm, and in the cat over 8 mm. It is subdivided into many compartments, each of which can trigger different emotionally tainted responses. These compartments are organized in a columnar structure. In the cat, part of the PAG elicits howling and hissing sounds as a sign of aggression, while another area elicits mewing, a sign of positive emotion (Subramanian et al., 2016, 2021).

In many species, including rodents, both escape and freezing responses can be elicited. The dorsomedial and dorsolateral PAG (together known as

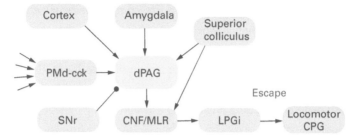

Figure 2.17
The escape reaction is coordinated from the dPAG. The dPAG elicits locomotion via the cunei-form nucleus (CNF), which in turn activates the lateral paragigantocellular nucleus (LPGi), which targets the spinal locomotor CPGs. The premammillary cck-expressing neurons in hypothalamus provides a major input to the dPAG, as does the cortex, the amygdala, and the medial parts of the SC.

dPAG) elicits escape behavior, while the ventrolateral PAG (vlPAG) elicits the freezing behavior. Both occur as response to fear, an immediate apparent threat leads to escape, while a more distant potential threat may lead to a freezing response after a risk assessment. There is a reciprocal inhibitory relation between the dPAG and vlPAG (Lefler et al., 2020). Escape and freezing cannot occur at the same time. The dPAG receives major input from the cholecystokinin (cck)–expressing neurons within the premammillary nucleus (PMd-cck), which in turn receives input from a variety of structures, including other hypothalamic nuclei and limbic structures (Comoli et al., 2000; Wang et al., 2021). Activation of the dPAG or PMd-cck reduces or blocks the escape response (figure 2.17). In addition, the dPAG receives input from the amygdala, superior colliculus (SC), and cortex (Lefler et al., 2020). Input from all areas that can trigger the escape reaction is thus mediated via dPAG. In parallel, there is also suppression of pain perception, which may function from the perspective that when an individual has escape as the priority, pain should be suppressed.

The lateral PAG (lPAG) mediates attack behavior and aggression. Optogenetic activation of the lPAG triggers attack behavior, and it receives dense projections from the ventromedial hypothalamus (VMH), a structure known to elicit attack behavior. Activation of either structure is efficient, and inhibition of the lPAG markedly reduces the effect of the VMH, and it is thought that the activation of the lPAG represents the motor aspect of aggression and the emotional aspect is conveyed via other projections of VMH (Lin et al., 2011; Falkner et al., 2014; Lo et al., 2019). As a medical student many years ago, I was shown a sweet and purring cat with an electrode implanted in its hypothalamus. When stimulated, there was a radical change, and the cat was transformed into a scary creature with hair and tail raised, looking around and then attacking a cork. When the stimulation ended, the cat sat down, looked around, and gave an impression of being surprised and wondering why it had acted as it did. After some minutes, it became again the friendly, purring cat that it was a pleasure to interact with. This was so impressive that more than 50 years later, I can recall the images, and perhaps the demonstration contributed to my interest in becoming a neurophysiologist.

Moreover, other compartments of the PAG can trigger lordosis behavior, part of the sexual behavior of female rats as studied by Pfaff and colleagues (e.g., Pfaff, 2017). Neurons in the VMH increase their level of activity when subjected to enhanced levels of estrogen. The VMH projects to neurons within the PAG that can mediate lordosis behavior via descending pathways from the lateral vestibular nucleus and medullary reticular formation, which in turn activate motor neurons that supply the dorsal trunk muscles. The lordosis reflex is triggered by sensory input from the skin in the pudendal region and adjacent skin areas that projects back to PAG via spinoreticular neurons and triggers the behavior. Only under the influence of VMH/estrogen will the PAG trigger a lordosis reflex as part of the female sexual act.

Yet another aspect of the PAG is its role in controlling the motor aspects of maternal behavior, such as pup grooming in mice. The behavior originates from GABAergic neurons in the medial preoptic area (MPOA) that enhance their activity during maternal behavior (Kohl et al., 2018). It projects to GABA-ergic interneurons within PAG. Activity in MPOA neurons will thus inhibit the GABA interneurons within PAG and thereby produce disinhibition of the glutamatergic PAG projection neurons responsible for mediating the downstream motor behavior. Blockade of the transmission from the MPOA to the PAG-projecting neurons reduces maternal pup grooming markedly.

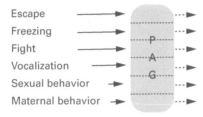

Figure 2.18
Different nuclei in the hypothalamus act on selected compartments of PAG, which in turn mediate the activation of downstream motor centers that elicit the patterns of behaviors shown here.

The PAG can thus issue commands for many forms of integrated motor patterns that affect respiration during vocalization, whole-body movements, such as escape or freezing, or specific reflexes to the dorsal trunk muscles, as during the lordosis behavior or maternal behavior (figure 2.18). It is, therefore, not surprising that different compartments of PAG receive specific input from a variety of structures, partially overlapping and partially specific. They include the GABAergic central nucleus of the amygdala, the inferior colliculus and SC, the prefrontal, visual, auditory, and motor cortices, the premammillary nucleus, the VMH, and a variety of other inputs. The medial habenula projects to the interpeduncular nucleus (IPN) and can also elicit freezing. The IPN projects to the *griseum centrale* (PAG) in lamprey and zebrafish (Olson et al., 2017; Cherng et al., 2020). On the output side, there is the cuneiform nucleus, which may trigger escape locomotion, the respiratory structures in the medulla, the parabrachial complex, the cervical spinal cord, and reticulospinal and vestibulospinal pathways.

A similar type of control center, located close to but not within the PAG, is the pontine micturition center (also called "Barrington's point"). It consists of a subgroup of glutamatergic neurons that project to micturition-related neurons in the sacral spinal cord. They express the corticotropin-releasing hormone, and when activated optogenetically, they elicit micturition (Hou et al., 2016). They receive input from the hypothalamus, olfactory bulb, cortex, brainstem nuclei, and spinal neurons affected by volume receptors from the bladder.

In conclusion, the PAG represents a conserved structure subdivided into compartments, each of which can trigger different integrated patterns of behavior, such as escape/freezing, aggression, specific vocalizations, the lordosis reflex, and maternal behavior, each with a distinct behavioral significance.

2.4.2 The Hypothalamus Controls Most Aspects of Behavior Important for Survival

As discussed in the preceding section, the hypothalamus controls most aspects of behavior that are of fundamental importance for survival. They include defense (freezing, escape, or fight), ingestive (fluid and food intake), and reproductive behavior, including maternal and paternal behavior, all of which represent innate circuitry that can be complemented with learned aspects. Moreover, the hypothalamus controls the endocrine system and regulates the diurnal rhythm.

The execution of these patterns of behavior or functions relies on projections to the PAG and downstream motor centers or glands. However, information concerning the downstream commands is also transmitted by ascending branches to the forebrain via the thalamus and a smaller projection directly to the striatum (Swanson, 2000). Many of the downstream commands are mediated via the PAG, which receives input from both the cortex and the SNr of the basal ganglia (McElvain et al., 2021; figures 2.17 and 2.18), and there is thus a possibility for the forebrain to control whether the hypothalamic commands will be transmitted to downstream motor centers or prevented to do so through gating.

The many subtypes of neurons of the medial hypothalamic cell columns are involved in the control of defensive, reproductive, and ingestive aspects of behavior. They comprise, from rostral to caudal, the medial preoptic nucleus, anterior hypothalamic nucleus, descending paraventricular, ventromedial and premammillary nuclei, and mammillary body.

The paraventricular descending nucleus is involved in the control of thirst and the initiation of drinking behavior. The regulation of water intake depends on the activation of hypothalamic osmoreceptors in the subfornical organ and of volume receptors in the circulatory system. Stimulation of this hypothalamic area leads to ingestion of water, as long as the stimulation continues, as first seen in the goat (Andersson, 1953). Moreover, the stimulation leads to a desire to obtain water, and the animal searches actively for water and is prepared to remove the lid of a bucket to obtain it (Andersson, 1953; Swanson, 2000; Zimmerman, 2020). The paraventricular nucleus projects to the PAG and a variety of downstream motor centers in the midbrain and brainstem and sends collaterals to the thalamus. To obtain water, there is a need to first activate the locomotor centers and then, when finding the water, position oneself in front of the bucket and finally activate the oral machinery to drink. The hypothalamic machinery thus activates an integrated neural machinery consisting of several motor circuits, acting one after the other, to satisfy the urge to obtain water.

The regulation of food intake is served by other circuits, primarily in the arcuate nucleus responsive to circulating agents that sense the need for food intake, such as leptin from adipose tissue and ghrelin from the gut, as well as information mediated via the vagal afferents and blood glucose levels. Hunger leads to foraging, which requires each species to search for its preferred food, remember the location of places in which it is likely to find food, explore and ultimately hunt for prey, or more peacefully find a location for grazing.

Another important aspect of life for most species is to identify a situation that is risky in terms of survival. For many species, the ability to suddenly freeze and remain motionless can be an effective strategy since many predators have difficulty to detect an object that does not move. We discussed in the previous section that neurons in the VMH project to a freezing compartment within the PAG that mediates freezing through downstream motor centers. If the predator comes too close, the strategy changes, and an escape at maximal speed is instead initiated via another part of PAG or, as the last alternative, an aggressive encounter occurs. For freezing/escape, two subtypes of interneurons in the VMH are engaged, one with estrogen receptors and another with steroidogenic factors (Kunwar et al., 2015; Lo et al., 2019; Stagkourakis et al., 2020; Falkner et al., 2020). Female sexual behavior is also mediated via estrogen-responsive neurons in the VMH, as discussed previously (Pfaff, 2017; Karigo et al., 2021).

The neuroendocrine control from the hypothalamus includes the release of oxytocin and vasopressin from the posterior part of the hypophysis and control of the release of hormones from its anterior lobe. This control system includes the paraventricular, anterior periventricular and arcuate nuclei, and part of the supraoptic nucleus (Swanson, 2000). They act either directly or via the adrenal gland, thyroid, and the ovaries or testicles. Clearly the release of hormones (e.g., sex and stress-related hormones) has a major impact on behavior. The diurnal rhythm, and thus the need to sleep during part of the day or night, is controlled from the suprachiasmatic nucleus.

The hypothalamus is a conserved structure throughout vertebrate phylogeny, and even though it only constitutes a limited part of the nervous system, it controls most aspects of behavior important for survival, including the endocrine system. But it does not operate in isolation, as its output can be controlled by the forebrain input to the PAG.

2.4.3 The Downstream Control of Behavior through the Amygdala: Fear Responses

The amygdala is considered as one main node for mediating fear responses in vertebrates from fish to humans (LeDoux, 2012; Janak and Tye, 2015; Adolphs & Anderson, 2018). The amygdala is not considered to mediate the subjective sensation of fear, which is thought to arise in other networks of the brain (LeDoux, 2012), but rather to mediate the efferent manifestations of fear, such as escape and freezing. Fear conditioning, as with an auditory cue followed by a painful stimulus, elicits after some trials a fear response when only the auditory cue is provided. These learned fear responses are mediated by the amygdala, as well as innate fear responses. The amygdala is a telencephalic nucleus with a large excitatory basolateral (BLA) part, which includes the lateral (LA), the basal, and the basomedial parts. It receives its major input from the sensory thalamus, the auditory and other sensory cortices, and the prefrontal cortex, all of which project to the LA. BLA in turn projects to the ventral striatum and prelimbic areas (Janak & Tye, 2015). In addition, BLA projects to the central nucleus (Ce) of the amygdala, which is GABAergic and acts on downstream centers in the midbrain and brainstem.

The fear responses are elicited through the PAG and various downstream centers and result in escape or freezing, autonomic effects on respiration, blood pressure and heart rate, as well as endocrine effects. The Ce acts through inhibition of GABAergic interneurons in PAG, which leads to a disinhibition of the PAG output neurons. Neurons in the ventrolateral part of PAG elicit the freezing response, and a more intense Ce activation can elicit escape reactions via the dorsolateral part of PAG. The effects are mediated via PAG neurons through projections to the medulla and then further to motoneurons (Tovote et al., 2016). The Ce neurons are divided into subtypes with specific molecular markers, which allows a more refined control of different aspects of behavior that are defensive as well as appetitive, although the details have yet to be elucidated. The basomedial nucleus of the amygdala acting via subtypes of neurons within the Ce can inhibit freezing responses (Adhikari et al., 2015) and is also thought to decrease the level of anxiety in mice. When this nucleus is activated, mice become bolder than in the control situation and walk freely on open surfaces that they would otherwise avoid (Tye et al., 2011).

2.5 INTEGRATION OF INNATE MOTOR PROGRAMS IN DAILY LIFE: SKILLED ASPECTS OF THE CONTROL OF MOTION

In the wild, most animals rely on the palette of innate motor programs available to them and are able to recruit them in an appropriate sequence for foraging, feeding, navigation, or escape and adapt them to the specific needs of the animal and to the surrounding environment (also discussed in chapter 5). It is important to realize that innate motor patterns (e.g., for locomotion) are flexible and modifiable. When considering the locomotor circuitry, for instance, one must consider not only the spinal CPGs with their sensory inputs, but also the corticospinal modulation for precision walking and other contributing structures as the cerebellum (Grillner & El Manira, 2020). One can walk forward, backward, or sideways, and at a low pace or running, while adapting to the environment. One can also learn new modifications of gait, such as walking with high heels, which is trouble-some at first, or imitating the walk of Charlie Chaplin (Grillner & Wallén, 2004). This is not only a human trait, but many mammals have this capacity. Consider, for instance, horses that can be trained to modify their gait in a precise, reproducible way for dressage competitions. Another example, the respiratory CPG ticks on at rest in an appropriate frequency and depth of breath, but it can be controlled at will. One can hold one's breath for many seconds or modify the pattern of breathing voluntarily. During speech, the respiratory pattern is modulated. In a large number of vertebrates, vocalization is produced by controlling the respiratory flow.

In tennis, the player runs rapidly toward the ball, combined with reaching out with the racket to hit it. This is essentially using the innate motor infrastructure for locomotion and reaching and adapt it to the specific dynamics of tennis. One clearly learns some aspects of playing tennis, such as interpreting the trajectory of the ball and placing the racket appropriately to send the ball back to a location where it cannot be reached by the opponent. With regard to the movement aspect of tennis, one learns to fine-tune the reaching movements and adapt them to the ever-changing situation that occurs with each shot. Another aspect of tennis is the serve, in which the situation is radically different because now the player is in control. A naive player has a large variability with each trial, whereas after training, the trajectory of the racket is almost exactly reproducible from trial to trial (Dhawale et al., 2017). This can be interpreted as fine-tuning of the innate reaching circuits to be able to gradually perform a dynamically reproducible movement. The basic motor infrastructure is available, but the timing of recruiting the specific aspects of the motor circuitry is a task in which the forebrain (and possibly the cerebellum) plays the crucial role.

2.5.1 Independent Finger Movements

A limited number of species, including humans, have the ability to control the movement of the individual fingers precisely and coordinate them in a delicate way, such as when playing the piano. All fingers are not equal, however; the thumb and index finger have the most versatility. One exquisite example of learned motor patterns is handwriting. To write individual letters, we have learned to coordinate mostly the thumb and index finger to control the movements of the pen and shape the letters. This is an example of how the innate circuitry to control the individual fingers is recruited in a precise way to produce a skilled motor pattern—that is, learning how to sequence the fingers in a precise way in relation to each other with separate remembered programs for each letter. We also have the possibility to, de novo, form entirely new movement patterns, such as when painting a scene on a canvas. This is based on our selective control of innate motor circuits for independent finger control.

The main message of this section is that humans have a very rich motor infrastructure at their disposal, extending from whole-body movements to independent finger movements in primates, and these building blocks can be recruited to design more complex movements. A skilled movement can be produced by combining distinct innate components in a well-timed sequence and tinkering with the amplitude of the various components, whether for playing tennis or grasping an object. These novel integrated motor patterns that can be perfected through training represent new skilled motor programs. They are put together by innate components from the motor infrastructure, and learned parts include the specific timing and fine-tuning.

During vertebrate evolution, there are species that essentially use their innate motor repertoire throughout life, learning to associate external signals with reward or threat and to adapt the motor patterns to external demands. In mammals, particularly primates, the motor apparatus becomes more versatile and available for learning new sequences, as in the impressive motor repertoire of the human hand.

2.6 CONCLUSION

The purpose of this chapter has been to illustrate the richness of various microcircuits and command structures that are available in the midbrain–brainstem–spinal cord that can be called upon by the forebrain, including the hypothalamus, to control most aspects of animals' behavioral repertoire, whether fish or primate. Different circuits are required for the execution of movements, generating rhythmic activity such as respiration, chewing and locomotion, circuits for swallowing, arm–hand coordination, vocalization, maternal behavior, and many other activities. Together, they form a motor infrastructure from which a great variety of movements can be recruited and integrated, resulting in purposeful and perfected motor sequences.

3 THE VERTEBRATE SOLUTION FOR ACTION IN THE EGOCENTRIC SPACE: MULTISENSORY INTEGRATION IN THE TECTUM/SUPERIOR COLLICULUS

3.1 INTRODUCTION

The tectum of the midbrain roof, in mammals called the "superior colliculus (SC)," is concerned with information from the immediate surrounding space—the egocentric space. This is mainly based on vision, but other senses, such as hearing, electrosensation in some fish, or the whiskers in rodents, also contribute to the processing in the tectum (Sparks, 1986). These circuits dynamically store information on where in the surrounding three-dimensional (3D) space a salient stimulus, a friend, a foe, or something edible is located, and can direct the gaze to the spot of interest and elicit orienting movements of head and body in the same direction. Conversely, if the object should be avoided, for instance to prevent a collision with a branch of a tree or a fellow pedestrian, an evasive movement can be elicited.

Movements in the egocentric space that depend on the dynamic input to the tectum/SC are controlled in a way that differs markedly from that of memory-based navigation depending on a maplike spatial memory required to find the remembered location of places for foraging or finding the way back to the nest. This type of memory utilizes the hippocampus and other parts of the brain. The latter type of navigation will be considered in chapters 4 and 5.

3.2 MULTISENSORY REPRESENTATION OF THE SURROUNDING SPACE IN THE TECTUM/SC

3.2.1 Retinotopic Map in the Tectum/SC

In all groups of vertebrates, vision is represented in a retinotopic manner in the most superficial layer of the tectum/SC. There is thus a map of the surrounding visual space in the tectum/SC (figure 3.2a). The information that originates from the retina is already preprocessed within the retina before it reaches the tectum/SC. In the zebrafish, there are at least 20 discrete categories of retinal ganglion cells that each forward different types of information of small edible objects or alarming stimuli in which the perceived object increases rapidly in size (see figure 3.1; Robies et al., 2014; Isa et al., 2021). The latter two types of retinal afferents terminate in the superficial layer of tectum but at somewhat different depths. Furthermore, the information forwarded to the tectal area corresponding to the anterior visual field may receive information about preylike objects and navigation, while those in the posterior visual field are more concerned with stimuli representing a potential threat.

In mammals, there are more than 30 subtypes of retinal ganglion cells that are defined functionally and by specific molecular or genetic markers (Shekhar & Sanes, 2021). In the lamprey, representing the oldest group of vertebrates, six subtypes of retinal ganglion cells have been identified morphologically, four of which project to the tectum with a differential projection pattern. The highest density is in the retinal area corresponding to the anterior visual space. The remaining two retinal ganglion cells project to the pretectum and are involved in the dorsal light response and visual escape reactions (Jones et al., 2009; Ullén et al., 1997).

How is the information processed in the retinotopic map (figure 3.2)? Salient stimuli will directly activate a limited set of excitatory neurons within the retinotopic map, and at the same time, they activate inhibitory interneurons (figure 3.2b; Kardamakis et al., 2015) that will inhibit other tectal neurons with a brief delay, locally over the entire visual map. Figure 3.2c shows that a local stimulation of the retina leads to a strong activation of only one part of the tectal map followed by inhibition, while in all other parts of the retina, only inhibition is evoked. If a distinct visual stimulus is recorded from somewhere

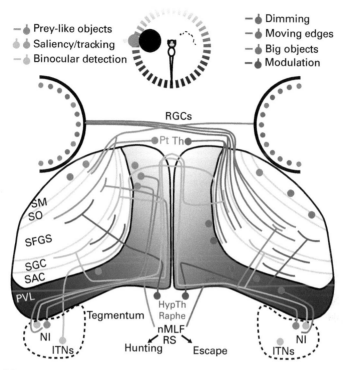

Figure 3.1

A schematic representation of the larval zebrafish's tectal inputs and outputs, illustrating the role of the tectum in receiving and integrating information from diverse sources. The inputs are strongly spatial—shading of a circle at the top and of the periventricular cellular layer (PVL) in the tectum—and differ depending on whether a small stimulus (prey seen by the right eye) or a large, looming stimulus (left eye) is presented. The processing of preylike stimuli (left tectum, since all retinal ganglion cell axons cross the midline) results in hunting behavior, while looming stimuli (right tectum) trigger escape responses. Inputs from various cell types and brain regions are color coded to indicate the stimulus properties that they encode (see legend at the top). Different types of information are delivered selectively to different laminae of the tectal neuropil. Abbreviations: HypTh, hypothalamus; ITNs, intertectal neurons; NI, nucleus isthmi; nMLF, nucleus of the medial longitudinal fasciculus; Pt, pretectum; PVL, periventricular cell layer; RGCs, retinal ganglion cells; RS, reticulospinal neurons; SAC, stratum album centrale; SFGS, stratum fibrosum et griseum superficiale; SGC, stratum griseum centrale; SM, stratum marginale; SO, stratum opticum; Th, thalamus. From Isa et al. (2021).

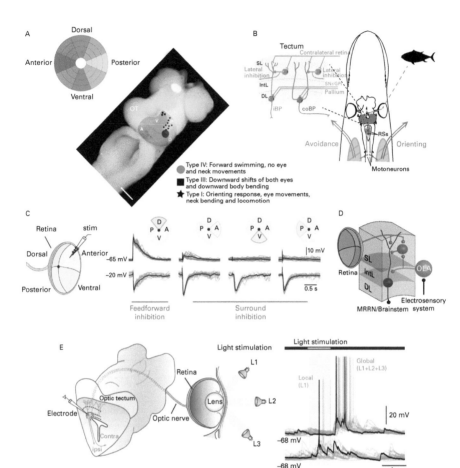

Type IV: Forward swimming, no eye and neck movements

Type III: Downward shifts of both eyes and downward body bending

Type I: Orienting response, eye movements, neck bending and locomotion

Figure 3.2

Structure and circuitry of the lamprey tectum. *A.* Color-coded schematic retina for the contralateral eye (left). Photograph of the lamprey brain (right), showing the location of the color-coded retinal input. The posterior retinal input terminates in the rostral tectum, the anterior retinal input terminates in the caudal tectum, and the ventral and dorsal retinal inputs terminate in the medial and lateral tectum, respectively. *B.* Drawing of the lamprey head and brain with the circuitry leading to orienting and avoidance responses. On the upper left, the tectal circuitry is represented, with the ipsilateral projecting neurons (iBPs) underlying avoidance reaction and the contralaterally projecting neurons (coBPs) responsible for orienting responses. These neurons receive monosynaptic input from retinal afferents and GABAergic interneurons. The output layer has input from both the pallium (cortex) and the substantia nigra pars reticulata (SNr). *C.* Stimulation of the retina in one quadrant leads to excitation in tectal neurons in the specific retinotopic projection area in the tectum, while stimulation in all other parts of the retina instead leads to a strong inhibition mediated by the tectal GABAergic interneurons. The red traces are recorded in the surrounding space, it will suppress other weaker stimuli that may reach the tectum simultaneously from other parts of the visual field. If two salient stimuli from two points of the retina arrive at the same time, there may be a rivalry, and the stronger one will take over. This scenario is illustrated in figure 3.2e. When one focal light source is activated, it leads to a strong activation of neurons in its target area (mauve trace). However, if other light sources are turned on at the same time, the excitatory response is entirely suppressed when the light is on, and after it is terminated, a postinhibitory rebound excitation occurs, as can be seen in the recorded traces. This circuitry in the tectum is thus designed to identify the location of salient stimuli in the egocentric space and direct the gaze and orienting movements to this area (as discussed next).

3.2.2 Visual and Electrosensory Afferents from the Same Point in Space Converge onto the Same Tectal Output Neuron: Multisensory Integration in the Lamprey Tectum

In the lamprey, visual information is transmitted monosynaptically from the retinal ganglion cells to the apical dendrites of the output cells, each of which has the cell body located in the deep tectal layer (figures 3.3a–c; Kardamakis et al., 2016). The dendrites of these cells transcend all layers of the tectum. The lamprey has cutaneous receptors that sense the electric

Figure 3.2 *(continued)*

with a holding potential of −65 mV, and the blue traces at −20 mV, when the inhibition can be seen clearly. There is thus a powerful lateral inhibition. *D.* Diagram illustrating the convergence between the retinal and the electrosensory inputs activated from the same point in space. At the distal dendrites, the retinal afferents form synapses, while the electrosensory afferents target the same dendrite, but closer to the cell soma. The tectal output neuron targets the middle rhombencephalic reticular nucleus (MRRN), located in the brainstem. *E.* Shows recordings from a lamprey eye–brain preparation in which the tectal output neurons can be patched, while the eye is illuminated with local brief light pulses (blue area) or globally with light in all parts of the retina. The neuron illustrated is excited by the local light (L1), while there is no response when the entire retina is illuminated (global). This illustrates the strong surround inhibition also shown in (B). There is, however, a postinhibitory rebound. Abbreviations: A, anterior; coBP, contralaterally projecting neuron; D, dorsal; DL, deep layer; iBP, ipsilaterally projecting neuron; IntL, intermediate layer; OLA, octavolateral area; P, posterior; RSs, reticulospinal neurons; SL, superficial layer; V, ventral. Modified from Jones et al. (2009) and Kardamakis et al. (2015).

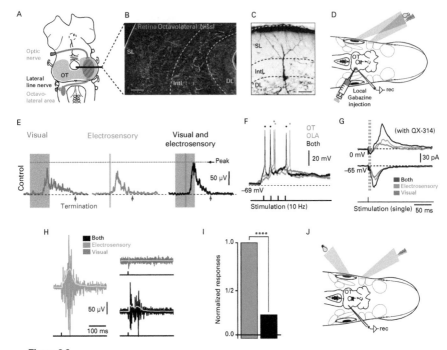

Figure 3.3

Visual and electrosensory signals summate in the deep layer of the lamprey optic tectum when originating from the same location in space and counteract if from different parts of space. *A.* Schematic of the lamprey brain showing the visual (blue) and electrosensory (red) afferents targeting the optic tectum (OT). *B.* Photomicrograph of the OT in a transversal view showing the retinal afferents reaching the most superficial layers (red), and the octavolateral (OLA) fibers innervating the intermediate layers (green). *C.* Morphology of an output neuron in the deep layer retrogradely labeled following a tracer injection in the middle rhombencephalic reticulospinal nucleus (MRRN) and filled intracellularly with neurobiotin while performing whole-cell recordings. Output cells extend their dendrites to the intermediate and superficial layers where the electrosensory and the visual inputs enter and terminate, respectively. *D.* Experimental settings for performing extracellular recordings during multisensory integration in the optic tectum. Dorsal view of the preparation, including the brain, eyes, and electrosensory areas (depicted by the skin patches), while driving output activity with light and electrical stimuli that are spatiotemporally aligned in the immediate surrounding. *E.* Rectified local field potentials obtained from visual, electrosensory, and bimodal sensory activation. *F.* Output neuron responses to sensory stimulation (train of 4 pulses at 10 Hz) of the optic tectum (blue traces), OLA (red traces), and bimodal (black traces) from rest at −69 mV. *G.* Inhibitory and excitatory postsynaptic currents (EPSCs) elicited by visual, electrosensory, and bimodal stimulation in an output cell recorded in a voltage clamp at 0 mV to show inhibitory currents, and at −65 mV (equilibrium for chloride-mediated GABAergic inhibition) to show excitatory currents. Drop lines show the onsets of excitatory and inhibitory currents. *H–J.* Spatially misaligned fields surrounding the head. Information from these receptors enters the brain via the octavolateral nerve and is transmitted to brainstem nuclei and then relayed to the tectum. The electrosensory information is also arranged in a spatial map calibrated to the visual map (figure 3.3b). If a visual stimulus is activated in a specific spot and an electrosensory stimulus is activated from the same spot, they facilitate each other (figure 3.3e), resulting in a more prominent activation. Also, at the level of the single neuron, the two stimuli summate (figures 3.3f and 3.3g), as demonstrated using both current clamp and voltage clamp. The excitatory currents from the visual and electrosensory traces summate (lower trace, black) and the inhibitory currents that have a longer latency (upper set of traces) also summate. However, if the two stimuli instead were elicited from different locations in space, they instead inhibit each other (see figures 3.3h and 3.3i). This means that visual and octavolateral stimuli from the same point in space facilitate each other, but from different points, they counteract each other.

When the visual and the electrosensory stimuli originate from the same spot in space, they will thus converge onto the very same output neuron. Whereas the visual afferents form synapses on distant dendrites, the electrosensory synapses are located closer to the cell body, still on the same dendrites (figure 3.3b), and the synaptic inputs from the two sources will summate. Thus, the signal from the two senses is processed by the same output neurons. From this, one can conclude that tectal neurons are designed to identify only the location of salient stimuli and do not care

Figure 3.3 *(continued)*

stimuli give rise to response reduction. We applied spatially disparate visual and electrosensory stimuli while recording responses in the contralateral optic tectum (*J*). Local field potentials in response to electrosensory stimulation (red trace) were drastically reduced when a different location of the tectal map was visually stimulated (blue trace). The responses when combining both sensory modalities are shown in black. The yellow traces on top show the rectified signals. Histogram showing the normalized responses for electrosensory activation (red), before and after simultaneously stimulating a visual off-region (black) (*I*). Abbreviations: OLA, octavolateral; OT, optic tectum; rec, extracellular recording electrode; SL, superficial layer; IntL, intermediate layer; DL, deep layer. Modified from Kardamakis et al. (2016).

about the sense from which the information originated, a true multisensory integration.

3.2.3 Auditory Contribution to Location in Egocentric Space: The Barn Owl Is the Champion

Sound travels comparatively slowly compared to light, and the difference between when the sound reaches the first and the second ear can be used to compute the location of the sound source, mostly in two dimensions. However, some species excel in this capacity, such as the barn owl. It can locate prey, such as a vole, from several meters or feet away with high precision, even when it is pitch dark, as reported in the classic studies of Mark Konishi and Eric Knudsen (Knudsen & Konishi, 1978; Knudsen, 1982; Carr & Konishi, 1990). This requires location in three dimensions, which is achieved by an asymmetric placement of the ears (with one ear somewhat higher than the other). In general, the information from the auditory hair cells in the cochlea is transmitted to the brainstem, and the processed information is then transmitted to a layer of the tectum/SC just below the visual layer. Here, a map of the surrounding space is formed based on auditory input, which is aligned with the visual spatial map. Vision and hearing complement each other regarding the spatial location, and they are tightly calibrated (Meredith & Stein, 1986; Stein & Stanford, 2008; Jay & Sparks, 1984). In mammals, the auditory information is also processed in the adjacent inferior colliculus.

3.2.4 Tectal Integration of Vision and the Heat-Sensing Pit Organ of the Rattlesnake

To identify the location of warm-blooded prey in the surrounding space, the rattlesnake and related species have evolved a heat-sensing pit organ on the head. It is innervated by heat-sensitive afferents in the trigeminal nerve, which via a relay in the brainstem are projecting to the tectum. Different thermosensitive afferents spatially arranged in the sensory epithelium of the pit can be activated, depending on where the heat source is located in the surrounding space. This information is projected to the tectum and represented in a maplike form. This heat map is in register with the visual map, and they complement each other (Hartline et al., 1978; Newman & Hartline, 1981). Individual tectal neurons receive excitatory input from both retinotectal and heat afferents. Again, location is what matters for the tectum, rather than which sense is the origin of the activation.

In conclusion, the tectum/SC is designed to locate salient stimuli in the surrounding space and adds the inputs from the available senses to identify the location of the object of interest. Multisensory integration occurs at the single-neuron level. The spatial maps can be formed by input from the retina, the cochlea, electrosensation, or thermosensation as discussed here, but somatosensation or input from the vibrissae can also take part.

3.3 THE TECTUM/SC CONTROL OF EYE, ORIENTING, AND EVASIVE MOVEMENTS

3.3.1 Orienting and Evasive Movements Elicited by Different Sets of Tectal/SC Neurons

The motor map in the deep output layer of the tectum/SC is aligned with the visual map in terms of spatial coordinates for the orienting movement. In the lamprey, these output neurons have a broad dendritic arbor, usually with two stem dendrites, which extend to the superficial layer where the axons of retinal ganglion cells terminate with synapses onto the dendrites, activating both α-amino-3-hydroxy-5-methyl-4-isoxazolepropionic acid (AMPA) and N-methyl-D-aspartate (NMDA) receptors. These synapses are located on tectal neurons receiving input predominantly from the anterior visual field, and they send their axons to the contralateral brainstem and make monosynaptic excitatory postsynaptic potentials (EPSPs) on reticulospinal neurons. An activation of this set of tectal neurons lead to an orienting movement of the head and body toward the visual stimulus (Kardamakis et al., 2015; Suzuki et al., 2019). The orienting movements are elicited by slowly expanding, nonthreatening stimuli.

In mice, the output layer of the tectum/SC consists of neurons, expressing the paired-like homeodomain transcription factor Pitx2, which are involved in the control of orienting movements of the head with different orientations in three dimensions. The neurons concerned with the orientation of the head in a specific direction are clustered in modules (Masullo et al., 2019; Wilson et al., 2018). The tectal/SC output layer is thus in reality composed of a large number of discrete modules, each specifying a certain head orientation, in register with the sensory maps in the superficial layer. Most likely, this organization is present in other vertebrate groups.

The other type of tectal output neuron is instead concerned with evasive movements, and it has only one stem dendrite that also extends to the visual layer. It is activated by fast-expanding, looming stimuli, which are generally perceived as threatening in most vertebrates. Their axons project to ipsilateral reticulospinal neurons, and, when activated, an evasive movement in the opposite direction will take place and the head and body will move away from the stimulus. In contrast to the orienting tectal neurons, the evasive tectal neurons are distributed over the entire visual field, which is to be expected since threatening stimuli can originate from any direction. It should also be appreciated that as an animal moves around in a complex terrain (e.g., a bird flying through the foliage of a tree, or a pedestrian walking on a busy street with others), it will have to make constant evasive corrections of the movement trajectory to avoid collisions. The ability to make fast evasive corrections is thus biologically a very important function for practically all living creatures.

In the lamprey, tectal orienting and evasive neurons have different membrane properties. The former has a lower threshold for activation by sensory stimuli than the evasive tectal neurons (Kardamakis et al., 2015; Suzuki et al., 2019). A weak sensory stimulus, like a slowly looming one, will then only activate orienting neurons but will be the subthreshold for the evasive neurons, and thus only an orienting movement will be elicited. A stronger sensory stimulus, however, will activate the evasive neurons and lead to activation of an evasive movement that will override the orienting response.

In rodents, the organization is similar, with the medial area of the SC representing the upper visual field eliciting freezing/escape and the anterior lower visual area representing the lateral tectum/SC area, from which orienting behavior can be elicited as needed, for instance during foraging (Dean et al., 1989; Sahibzada et al., 1986).

3.3.2 Eye Movement and Eye-Head Coordination

In the lamprey tectum, low-threshold stimulation in the area corresponding to the anterior visual field elicits coordinated eye movements, so that if the ipsilateral eye moves forward, the contralateral eye moves backward, and vice versa. With a somewhat larger stimulus strength, orienting movements and often locomotor movements will be induced (Saitoh et al., 2007). In the anterior part of this tectal area, small-amplitude movements are elicited, while more caudally, the eye movements become larger. The output from the tectum/SC is most likely conveyed via the horizontal and vertical gaze center in the brainstem and midbrain, respectively. The basic organization of the eye motor nuclei and the different eye muscles are conserved from the lamprey to primates (Fritzsch et al., 1990). In a caudal tectal area, locomotion is generated without eye and orienting movements, and in an area close to the caudal midline, large-amplitude struggling movements can be elicited, accompanied by large-amplitude synchronized eye movements.

These results from the lamprey agree with those from other vertebrate groups. In primates with the eyes oriented forward, an additional group of neurons have been identified that become activated during reaching toward objects in the surrounding environment (Philipp & Hoffman, 2014; Cooper & McPeek, 2021). These neurons are located in the lateral posterior part of the SC in an area related to the anterior visual field and can possibly be regarded in the context of orienting movements. This type of neurons has not been identified in other vertebrates, but that does not necessarily mean that they do not exist.

3.3.3 Visuomotor Coordination through the Tecto/SC-Thalamo-Cortical Link: Blindsight

In all vertebrates, the cortex and the pallium, its counterpart in lower vertebrates, project to the tectum/SC. Motor areas, such as the frontal eye field (FEF) in the frontal lobe, have monosynaptic projections to the tectal output neurons and can thus utilize the tectal circuitry to initiate movements of the eye/head from the cortex. The cortex may also facilitate subthreshold sensory stimuli from the retina or other senses and thereby initiate movement. In mammals and other groups, however, there are also projections from the visual areas in the cortex that target the sensory layers in the tectum/SC.

In addition to the motor commands issued from the tectum/SC to motor centers, a very extensive projection from the tectum/SC targets neurons in the thalamus (in mammals, the pulvinar nucleus), with further projections to the higher-order visual areas that are important for visuomotor coordination.

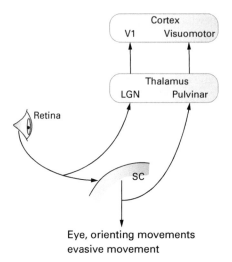

Figure 3.4
Interactions between the retina and the superior colliculus, thalamus, and cortex. Abbreviations: LGN, lateral geniculate nucleus; SC, superior colliculus; V1, primary visual area.

The pulvinar is separate from the lateral geniculate nucleus (LGN), which sends information to the primary visual area (V1). The information conveyed from the tectum/SC to the cortex can be regarded as preprocessed visuomotor data related to the ongoing activity in the tectum/SC and the commands issued to different motor centers from the tectum/SC (see figure 3.4). Importantly, tectal output neurons that target the brainstem may also have axonal projections directed to the thalamus, testifying that they mediate efference copy information to the thalamus and then the cortex (Capantini et al., 2017).

In humans, other primates, and other mammals, lesions of the V1 cause blindness in terms of object recognition. Individuals with such lesions can, although not consciously, become aware of perceiving an object like a ball being thrown toward them and then grab the ball or avoid being hit. This is called "blindsight" (Weiskrantz et al., 1974; Kato et al., 2011; Isa et al., 2021). This finding, and other pieces of evidence indicate that this tecto-thalamo-parietal projection is of critical importance for all forms of visuo-motor coordination. Even a complex interaction with dynamically moving objects in the environment can thus be handled without a conscious perception of the moving object, presumably through the tecto-thalamic pathway to the cortex (see chapter 5). This most likely suggests that the tectum/SC processing is used extensively for the planning of visuomotor commands at the cortical level, even when V1 is intact (Beltramo & Scanziani, 2019; see also chapter 5).

3.3.4 Salient Stimuli in the Tectum/SC Activate the Dopaminergic Neurons in the SNc

In both lampreys and mammals, there are projections from the tectum/SC to the SNc. Figures 3.5a and 3.5b show that both visual and electrosensory stimuli activate dopamine neurons in the SNc, and the graph in figure 3.5c shows that the more intense the stimulus is, the stronger the activation of SNc becomes. The salient stimuli that activate the tectum/SC are transmitted to the SNc and leads to a burst of activity in the dopamine neurons (Redgrave & Gurney 2006; Pérez-Fernández et al., 2017), which project not only to the striatum, but also to the mesencephalic locomotor region (MLR)

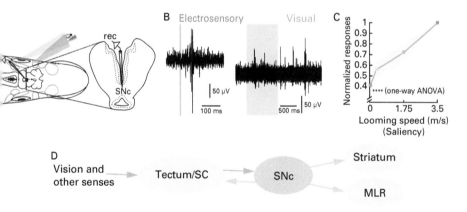

Figure 3.5
SNc respond to sensory stimuli. *A.* Schematic of the preparation used for applying visual (blue) and electrosensory (violet) stimuli while recording SNc activity (left), which was reached through the ventricle, as shown in the transverse section (right). *B.* Local field potentials were recorded in the SNc in response to both brief electrosensory stimulation (left) and flashes of light (right). *C.* Plot illustrating the increase of SNc activity in parallel with increasing looming expansion speeds (shown here as increases in diameter), as reflected by the amplitudes normalized to the maximal response (mean ± SEM). *D.* Diagram illustration of how sensory stimuli reach the SNc. Abbreviations: MLR, mesencephalic locomotor region; rec, recording; SC, superior colliculus; SNc, substantia nigra pars compacta. Modified from Pérez-Fernández et al. (2017).

and back to the tectum/SC (Comoli et al., 2003; Ryczko et al., 2016; Pérez-Fernández et al., 2014, 2017; see the diagram in figure 3.5d). This organization is characteristic of all the vertebrates that have been investigated.

Dopamine activity often precedes a spontaneously initiated movement (da Silva et al., 2018) and, by acting on D1 dopamine receptors (D1Rs), it will facilitate movement through an action on the striatum, but also by lowering the threshold for activation of the MLR (Ryczko et al., 2016). With regard to the action on the tectum/SC, the output neurons that express D1Rs in the tectum/SC are facilitated, and thereby the visuomotor reflexes to eye muscles are also facilitated. Salient stimuli will thus, if sufficiently prominent, activate dopamine neurons, which in turn will facilitate the processing within the tectal microcircuitry (see Pérez-Fernández et al., 2017).

3.3.5 Projections from Different Areas of the Cortex to Separate Zones in the SC

A recent study in the mouse showed that the SC can be divided into four zones based on the input from different parts of the cortex (Benavidez et al., 2021; see figure 3.6). The most lateral zone receives input from the somatosensory and motor cortices, while the centrolateral area receives input from gaze-related cortical areas such as the FEF, with the axons terminating directly on the output neurons of the SC. These SC areas also receive input from retinal afferents activated from the anterior visual field, from which orienting moments are activated in other species. The output from these two SC zones is related to eye-head coordination and orienting behaviors, such as foraging.

The medial and centromedial areas of the SC are related to the upper visual field and are thought to process potentially threatening stimuli and to mediate defensive or aggressive behavior. The medial SC area receives input from the V1 that targets the superficial layers of the SC, while the centromedial area receives input from the higher-order visual areas and the ventromedial hypothalamus.

Figure 3.6 also displays the projection pattern from the cortex to the striatum, showing that the dorsolateral and ventral part of the striatum corresponds to that received by the two lateral SC areas, while the dorsomedial striatum receives cortical input from the same areas as the two medial SC areas. This organization is also conserved in the output nucleus of the basal ganglia,

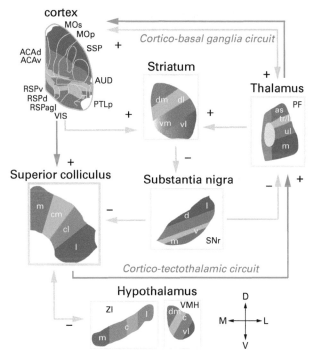

Figure 3.6

Topographic organization of the connectivity of the mouse SC. The four zones of the SC are connected either unidirectionally or bidirectionally with the cortex, parafascicular thalamic nucleus (PF), zona incerta (ZI), ventromedial hypothalamus (VMH), and SNr. Plus and minus signs indicate excitatory or inhibitory projections, respectively. A directional compass can be seen at the bottom right. Redrawn and modified from Benavidez et al. (2021).

substantia nigra pars reticulata (SNr), in the thalamus, and even with regard to input from the hypothalamus. From figure 3.6, it is clear that the forebrain and SC are organized in one set of areas related to the anterior lower visual field (blue-green in figure 3.6) that control such movements as foraging and gazing. On the other hand, the upper visual field is represented in connected areas from the cortex to the medial SC and is focused on detecting potential threats and responding by escape or aggression (yellow-red color).

3.3.6 The Basal Ganglia Control the Output from the Tectum/SC by Releasing Movements through Disinhibition

As first described by Hikosaka and Wurtz (1983), the SNr contains tonically active GABAergic neurons, which under resting conditions inhibit all tectal/

SC output neurons. The basal ganglia will thus be able to prevent any outgoing actions from the tectum/SC. It is only when the SNr neurons become inhibited from the striatum that tectal/SC neurons become disinhibited and ready to send out commands for eye, orienting, or evasive movements (see chapter 4).

The neurons of the SNr are organized so that the various parts of the SC are controlled separately (McElvain et al., 2021), with the most lateral SNr neurons projecting to the lateral part of SC, and those targeting the medial SC originating from the medial SNr. Moreover, the cellular properties of the SNr neurons projecting to the various parts of the SC differ in terms of rest rate and the shape of the action potential.

3.4 CONCLUSION

The tectum/SC contains multisensory spatial maps that detect where a salient stimulus is located within the surrounding space and use the aligned motor maps to elicit saccadic and orienting head-neck movements toward the salient points. The anterior lower visual field is concerned with orienting movements, while the upper visual space is concerned with potentially threatening stimuli and can trigger evasive movements and defensive or aggressive patterns of behavior. The rodent SC can be subdivided into four areas; the two lateral ones deal with the control of gaze and orienting movements, whereas the two medial ones are concerned with defense and aggression. Each of these areas receives specific projections from both the neocortex and the SNr. In addition to motor actions, the SC forwards processed information to the thalamus that is further relayed to sensory visuomotor areas in the parietal lobe. Salient stimuli in the SC also activate dopamine neurons in the SNc. The tectum/SC is evolutionarily conserved from the lamprey to mammals and plays a central role in the control of behavior in all vertebrates.

4 THE ROLES OF THE BASAL GANGLIA: FOR INITIATION OF MOVEMENT AND MOTOR LEARNING

Nothing happens until something moves.

—ALBERT EINSTEIN

4.1 OVERVIEW: THE RELATION BETWEEN THE CORTEX AND THE BASAL GANGLIA

In the previous chapters, we have dealt with the various circuits in the midbrain, brainstem, and spinal cord that can execute the many aspects of the complete integrated motor behavior of the different vertebrates. With the analogy used earlier, they correspond to the members of an orchestra playing different instruments, or the keys of a piano. We now move on to consider the forebrain, which takes the role of the conductor of the orchestra or the pianist pressing the keys of the piano.

Let us once more remember the quote from C. S. Sherrington: "All the brain can do is to move things, whether whispering or felling a forest," and thereby the fundamental role of the motor systems. The forebrain is in control of which microcircuits to activate in a given situation, when playing tennis, walking, or just talking.

The forebrain includes the cerebral cortex (which is discussed further in chapter 5), the basal ganglia including the dopamine system, and the thalamus conveying information from our senses regarding what happens around us informing both the cortex and the basal ganglia (figure 4.1). The hypothalamus (discussed already in chapter 2) can also be considered as

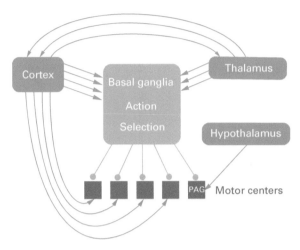

Figure 4.1
Control of motor behavior. The basal ganglia receive input from the cortex and thalamus and target various motor centers. The cortex also receives input from the thalamus and sends information directly to the motor centers (dark blue). The control of escape behavior is through projections from the hypothalamus to the PAG.

part of the forebrain, as it plays an important role in the control of instinctive behaviors such as escape reactions, aggression, and foraging and several autonomic and endocrine circuits that affect behavior. The hypothalamic nuclei also utilize the microcircuit keyboard of downstream motor centers, including the periaqueductal gray (PAG).

The motor centers in the midbrain, brainstem, and spinal cord are directly or indirectly under tonic inhibitory control from the basal ganglia output nuclei (figure 4.1) under resting conditions. Consequently, all these centers will be difficult to activate unless they are relieved from this inhibition. The evolutionary rationale for this is presumably that it is important to ascertain that the motor circuits will remain inactive until called into action in order to avoid conflicting motor programs being recruited at the same time. A removal of the inhibition (known as "disinhibition") from the basal ganglia is generally required to activate a specific motor center. The basal ganglia circuits operate based on information received from the cortex, thalamus, and dopamine and serotonin systems in order to determine whether a motor center should be selected for action. The disinhibition from the basal ganglia may be complemented by direct excitation from the cortex to specific motor centers (figure 4.1).

The relative role of the cortex and the basal ganglia in different behavioral contexts is somewhat ambiguous and will vary under different conditions. Arber and Costa (2018) suggested that cortex would broadcast a wish, while the basal ganglia would determine whether the wish may be fulfilled.

In mammals, such as rodents, rabbits, and cats, removal or inactivation of the neocortex in young animals with the rest of the brain left intact has surprisingly little effect on the standard motor repertoire. Such "decorticated animals" will walk around, perform exploratory movements, search for food and water, eat, and go through phases of sleep and active behavior (figure 4.2; see also Bjursten et al., 1976). They can live for years in a laboratory environment. This means that the process of initiation of patterns of behavior adapted to the needs of the animal will be handled primarily by the basal ganglia under these conditions with input from the thalamus,

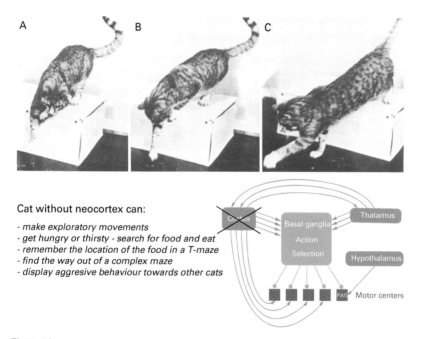

Cat without neocortex can:

- *make exploratory movements*
- *get hungry or thirsty - search for food and eat*
- *remember the location of the food in a T-maze*
- *find the way out of a complex maze*
- *display aggresive behaviour towards other cats*

Figure 4.2
A cat without the neocortex. From Bjursten et al. (1976).

but not the cortex. Basic drives from hypothalamic circuits to the PAG and downstream motor centers may also contribute to actions such as foraging. However, without the cortex, they will have difficulty in interpreting signals from the surrounding world and often react with "sham rage" and attack without obvious reason.

The Ölveczky laboratory (Kawai et al., 2015) showed that a learned behavior such as a double lever press at a fixed time interval, a demanding task for a rat, could be produced with the same precision after removal of the frontal motor areas. The motor cortex was thus not needed for the initiation and execution of this well-learned motor sequence. However, the motor cortex was required during the learning period; if it were removed before the training, the rat could not learn the task. To perform the task, the thalamic input to the striatum was required and the dorsolateral part of the striatum had to remain intact, whereas the dorsomedial part of the striatum was not required. There is thus an intricate interaction between the cortex and striatum in the learning phase, and these parts of the brain are required for the ability to learn the task.

Transection of the corticospinal axons in monkeys, leaving the rest of the cortical control of the basal ganglia, midbrain, and brainstem intact, leads to a loss of independent finger movements, but all other parts of the motor repertoire remain largely intact (Porter, 1987; Lawrence & Kuypers, 1968a,b). However, dynamic tasks requiring visual interaction with moving objects or fine manipulation of objects may be more dependent on processing within the cortex and input from the retina via the superior colliculus (SC). Also, precision movements with independent finger movements in primates, such as playing the piano or knitting, may depend on the direct corticospinal projections to motoneurons.

Hypothalamic nuclei act directly via the PAG to elicit escape, freezing, and defense reactions (see figure 4.1 and figure 7.1 in chapter 7). These are reactions that require a rapid response for survival, and therefore may initially bypass the controlling action of the basal ganglia and cortex. In this chapter, we will discuss the details of the basal ganglia and their interaction with other parts of the forebrain and downstream motor centers.

4.1.1 Specific Tasks for the Forebrain in the Control of Motion

It is important to consider the main tasks of the forebrain in the context of controlling our movement repertoire. They can be summarized as follows:

- To select, initiate, and terminate movements that we know well, such as saccadic eye movements, walking, or reaching for an object. The motor programs for these movements are readily available and do not need learning. Nevertheless, there is always a need for calibration to changing body dimensions and adaptation to the surrounding environment.
- Another factor that needs to be considered is the amplitude of the movement in a movement such as reaching, and the definition of the target (e.g., for a pianist, whether to press the keys gently or forcefully).
- Learned movements like habits, in which a motor sequence is stored, can be recruited whenever needed. This can be the simple cases, such as how to handle the keys when opening the door to your home, or more complex sequences, which often combine a set of innate programs into a habit.
- The process of learning new sequences, such as hitting a tennis serve. The initial trials are quite variable with little success, and after training, a very reproducible motor pattern can be produced. Similarly, when children learn to write, they form each letter with high variability at first, but after a learning process, a distinct shape for each letter is formed. As Bernstein (1967) was the first to point out, the overall shape of an "A" remains very similar regardless of whether it is written with a pencil or on the blackboard, although the dimensions can differ with orders of magnitude, and completely different sets of muscles are used to form the letter.

In this chapter, we will discuss the processing in the basal ganglia that underlies the control of motion on the cellular, network, and systems levels. We will focus initially on the mechanisms that allow selection, initiation, and termination of movement.

4.2 BASAL GANGLIA: ORGANIZATION

The basal ganglia are evolutionarily conserved and had evolved in the oldest group of vertebrates, the lamprey, to control basic aspects of motor behavior

(Grillner & Robertson, 2016; Grillner, 2021). The basal ganglia play a major role in controlling both the initiation of movement and for learning new patterns of coordination. They operate through the control of genetically defined circuits in the midbrain, brainstem, and (indirectly) the spinal cord. Copies of the information sent to the brainstem motor centers are also reported back to the cortex via the thalamic nuclei. The main functions can be summarized as follows:

- The basal ganglia contribute to the release of various motor acts through disinhibition, such as saccadic eye movements.
- The basal ganglia take part in the selection between the many motor programs/microcircuits at the level of the striatum, leading to the release of a specific motor behavior.
- The basal ganglia are thought to control the amplitude of a selected movement (i.e., should you hit the piano key lightly or with full strength, as in a Stravinsky piece).
- The basal ganglia allow one to learn, store, and recall new motor sequences, such as playing a piece on the piano, or the more trivial motor sequences used in everyday life, such as tying your shoelaces. Here, one compares habits stored in a long-term perspective and more flexible, goal-directed patterns of behavior.
- Dysfunction of the basal ganglia can lead to hypokinesia, as in Parkinson's disease, with difficulty to initiate movements or perform complex motor sequences, or conversely, hyperkinesia with excessive movement initiation, as in Huntington's disease or L-DOPA- induced dyskinesia.

Next, we will review how the basal ganglia perform their many roles, what is known, and what remains enigmatic or unknown to date. We will first discuss the structure and function of the circuits within the basal ganglia. Subsequently, we will integrate this information and consider how the many functions of the basal ganglia have been solved, or what is currently the most reasonable interpretation of the basic mode of interpretation of these roles.

An overview of the human and rodent basal ganglia with the various subnuclei is shown in figure 4.3. The striatum, the input stage of the basal ganglia, receives afferents from the cortex and thalamus. In primates, it is

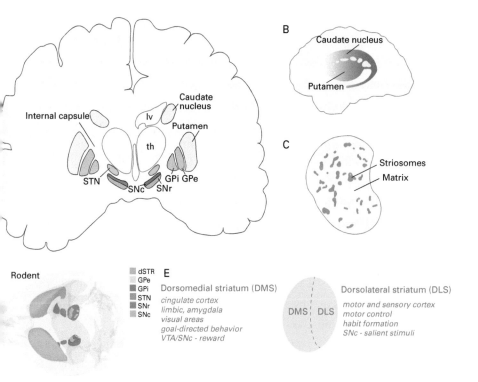

Figure 4.3

The basal ganglia subnuclei in the human and rodent brain. *A.* The location of the basal ganglia subnuclei at the level of the thalamus. *B.* A sagittal view of the brain showing the shape of the caudate-putamen. *C.* Schematic of the striatum, indicating the matrix and striosome compartments. *D.* Horizontal view of the rodent brain, indicating the subnuclei shown in relative sizes. *E.* Subdivision of the dorsomedial striatum (DMS) and dorsolateral striatum (DLS) and the functions ascribed to them, shown schematically. Abbreviations: dSTR, dorsal striatum; GPe, globus pallidus externa; GPi, globus pallidus interna; lv, lateral ventricle; SNc, substantia nigra pars compacta; SNr, substantia nigra pars reticulata; STN, subthalamic nucleus; VTA, ventral tegmental area.

subdivided into the caudate nucleus and the putamen, which corresponds to the rodent dorsal striatum. In rodents, the axons from the cortex that run via the capsula interna to the brainstem do not subdivide the striatum into two discrete parts, the caudate and putamen. The dorsal striatum in rodents is mostly subdivided into dorsolateral (DLS) and dorsomedial (DMS) parts with different inputs and functions. The DLS is often referred to as the "somatomotor striatum," with input from the motor and somatosensory parts of the cortex, and the DMS as the associate branch of the striatum, with input from associate areas of the cortex (figures 4.3d and 4.3e). The most caudal part of the dorsal striatum is now often considered a separate element, referred to as the "tail striatum" (Valjent & Gangarossa, 2021), with input from the visual and auditory cortex and amygdala (Reig & Silberberg, 2014; Znamenskiy & Zador, 2013). There is also the ventral striatum, which is smaller than the dorsal striatum and is comprised of the nucleus accumbens and the olfactory tubercle, with input from the limbic and olfactory systems, respectively.

Furthermore, the striatum is subdivided into a larger matrix volume and a much smaller striosomal, "patchy" part (figure 4.3c)—these two compartments have different functions (Graybiel & Ragsdale, 1978). The matrix is involved in the selection of action and the control of motion and behavior in general, whereas the striosomal part controls the level of activity in the dopamine neurons in substantia nigra pars compacta (SNc), and presumably an evaluation of action (as discussed next).

The striatum is a large structure. In the mouse, it contains 850,000 GABAergic neurons and in the rat there are 1,350,000 neurons (Boyes & Bolam, 2007) on one side of the brain, out of which 95 percent are striatal projection neurons (SPNs). The SPNs are subdivided into one half (dSPNs; direct pathway striatal projection neurons) projecting directly to the output compartment, substantia nigra pars reticulata (SNr; direct pathway), and the much smaller globus pallidus interna (GPi; direct pathway); and the other half is the globus pallidus externa (GPe; indirect pathway), indirect pathway striatal projection neurons (iSPNs). In addition, 5 percent of the interneurons can exert a major effect on the operation of the striatum (figure 4.4a). There are three major subtypes: cholinergic interneurons (ChINs; orange color), fast-spiking interneurons (FS; green), and low-threshold-spiking interneurons (LTS; mauve). In addition, there are several other smaller groups with less well known functions, such as the tyrosine hydroxylase-expressing (TH) interneurons, 5-HT3-expressing types, and neurogliaform types (gray area in figure 4.4b). The connectivity between SPNs and the three interneuron subtypes is outlined in figure 4.4c.

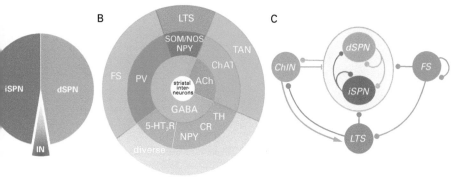

Figure 4.4

Striatal interneurons and the striatal microcircuit. *A.* Diagram showing the relative proportions of indirect and direct SPNs and interneurons. *B.* Each subtype of striatal interneurons identified by their neurotransmitter expression (inner circle), other molecular markers (middle circle), and electro-physiological properties (outer circle) are represented in the circular plot. The group shown in gray is a diverse group here labeled as "diverse." Redrawn and modified, with permission, from Burke et al. (2017). *C.* The striatal microcircuit with the connectivity between the SPNs and their input from FS, LTS, and ChIN interneurons. Abbreviations: ACh, acetylcholine; ChAT, choline-acetyl transferase; ChIN, cholinergic interneuron; CR, calretinin; dSPN, direct pathway striatal projection neuron; FS, fast-spiking interneuron; GABA, γ-aminobutyric acid; 5-HT3R, serotonin type-3 receptor; IN, interneurons iSPN, indirect pathway striatal projection neuron; LTS, low-threshold-spiking inter-neuron; NOS, nitric oxide synthase; NPY, neuropeptide Y; PV, parvalbumin; SOM, somatostatin; SPN, striatal projection neuron; TAN, tonically active neuron; TH, tyrosine hydroxylase.

4.2.1   Interaction within the Basal Ganglia:
Direct and Indirect Pathways

The interactions among the main compartments of the basal ganglia are summarized in figure 4.5, providing a functional scheme. The output stage, the SNr and GPi, consists of spontaneously active GABAergic neurons that receive inhibitory input directly from the striatum via dSPNs and is therefore referred to as the "direct pathway." The axons of the SNr/GPi in turn project to motor centers in the midbrain and brainstem and send collaterals to the thalamus that provide information to the cortex and striatum. The direct pathway dSPNs express dopamine D1 receptors and substance P and inhibit the spontaneously active SNr/GPi neurons, thereby disinhibiting a specific motor center and facilitating the initiation of movement.

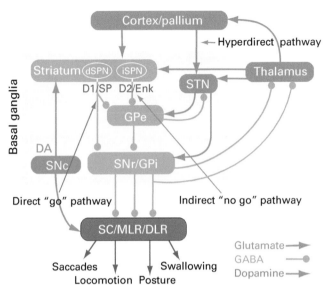

Figure 4.5

The organization of the basal ganglia. The striatum consists of GABAergic neurons, as do the GPe, GPi, and SNr. The SNr and GPi represent the output level of the basal ganglia and project via different subpopulations of neurons to the superior colliculus (SC), the mesencephalic (MLR), and diencephalic (DLR) locomotor command regions (as previously discussed) and other brainstem motor centers, as well as back to the thalamus with efference copies of information sent to the brainstem. The dSPNs that target the SNr/GPi express the dopamine D1 receptor (D1) and substance P (SP), while the iSPNs express the dopamine D2 receptor (D2) and enkephalin (Enk). The indirect loop is represented by the GPe, STN, and output level (SNr/GPi)—the net effect being an enhancement of activity in these nuclei. Also indicated is the dopamine input from the SNc (green) to the striatum and brainstem centers.

The indirect pathway has an opposite effect of that of the direct pathway. It originates with the iSPNs, which inhibit spontaneously active GPe neurons (the prototypic subtype) that in turn target neurons in the SNr/GPi. It thus reaches the SNr/GPi indirectly (therefore the name). The net result is that the iSPN inhibition of the spontaneously active GPe neurons will lead to a disinhibition of the SNr/GPi and will thus produce enhanced activity, and thereby further inhibition of the downstream motor centers, an effect opposite to that of the direct pathway. In addition, iSPN-induced inhibition of GPe will reduce the inhibition of the subthalamic neurons (STNs) and thereby cause enhanced activity of the glutamatergic STNs that enhances the activity of the SNr/GPi. The STN also receives excitation from the cortex and thalamus, referred to as the "hyperdirect pathway," which will act to terminate motor activity by exciting SNr/GPi neurons, as will the indirect pathway (see also figure 4.16).

The striatum also receives a strong dopamine innervation from the SNc, whereas the other subnuclei within the basal ganglia receive less dense dopamine input. Further, dSPNs express excitatory D1 receptors, whereas iSPNs instead have inhibitory D2 receptors and thus the direct and indirect pathways will be affected in opposite ways by increased dopamine activity. In addition, the SNc targets the downstream motor centers and can directly influence the excitability of these centers.

The dopamine system consists of the SNc, which targets the DLS, while the adjacent ventral tegmental area (VTA) targets the DMS and ventral striatum (accumbens). The striatum is much larger than the other basal ganglia nuclei. The GPe has only 1.7 percent of the neurons in the striatum, the SNr/GPi contains 1 percent, STN 0.5 percent, and SNc 0.3 percent (rat; Oorschot, 1996). The exact proportions have not been described for other groups, but the relative differences between the striatum and other structures most likely apply. The organization of the basal ganglia is evolutionarily conserved, and the scheme described here applies to both mammals and lampreys, and and perhaps all other vertebrates as well (Grillner & Robertson, 2016).

4.2.2 The SNr/GPi: The Output Level of the Basal Ganglia Controls Downstream Motor Centers and Provides an Efference Copy Back to the Cortex and Striatum

We will first consider the output targets of the basal ganglia in the midbrain, brainstem, and thalamus. A characteristic of the SNr/GPi is that the GABAergic neurons are spontaneously active under resting conditions and thereby continuously inhibit their downstream motor centers. The evolutionary rationale for this is presumably to ascertain that motor centers should remain inactive until they are specifically called into action. A recent study in the mouse has shown that the SNr targets no fewer than 42 downstream structures in the midbrain and brainstem (McElvain et al., 2021) and is presumably able to control many of them individually. By selectively inhibiting a subpopulation of SNr/GPi neurons involved in the control of a specific motor program, the downstream motor center is released from inhibition and then recruited to action. This disinhibition can be complemented by excitation from other sources, such as the cortex.

McElvain et al. (2021) extended earlier findings from the cat and primates showing that SNr neurons affect the control of saccadic eye movements generated from the SC, the control of locomotion and swallowing, and other brainstem centers (Takakusaki, 2008; Hikosaka & Wurtz, 1983: Grillner et al., 2005). Figure 4.6 shows some of the specific targets of the SNr and the various compartments such as the lateral SC, which affect the orienting movements of head and body and the related eye movements, while the medial part relates to threatening stimuli and elicit escape and defense reactions mediated via PAG. The motor centers in the medullary and pontine reticular formation control orofacial movements, reaching, grasping, hindlimb movements, and muscle tone. The SNr is thus in the position to control independently the activity in each of these motor centers and to determine whether to remain inhibited or be called into action. These motor centers from the midbrain to the spinal cord represent a large part of the motor repertoire of most vertebrates, which shows how central the SNr is to the control of movements.

McElvain et al. (2021) also showed that all SNr neurons that target downstream motor centers have a collateral to different portions of the thalamus, the motor thalamus, and the parts of the parafascicular (PF) nucleus within the intralaminar thalamic nuclei (figures 4.6 and 4.7). The latter projects back to the striatum and the cortex. Subpopulations of SNr project to compartments within the PF, which in turn target separate parts of the striatum. This means that both the striatum and cortex receive specific information about the commands sent to downstream motor targets. The collateral information sent via the thalamus to the cortex and striatum provides what is often referred to as an "efference copy" or a corollary discharge. This information about current motor commands to downstream motor centers is naturally of critical importance for planning the next phase of the movement. In much of the

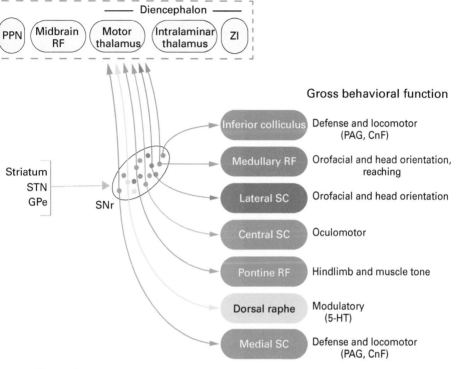

Common collateral targets

Gross behavioral function

Target	Function
Inferior colliculus	Defense and locomotor (PAG, CnF)
Medullary RF	Orofacial and head orientation, reaching
Lateral SC	Orofacial and head orientation
Central SC	Oculomotor
Pontine RF	Hindlimb and muscle tone
Dorsal raphe	Modulatory (5-HT)
Medial SC	Defense and locomotor (PAG, CnF)

Figure 4.6
Summary of SNr output pathways. Each SNr population projects to functionally distinct brainstem regions and send collaterals to brainstem nuclei and a set of common target regions. All the populations of projections demonstrate a distinct one-to-many projection pattern. Redrawn and modified from McElvain et al. (2021).

current literature, the loop back from the SNr to the cortex is still viewed as if the ultimate motor command would be issued from the cortex, but to a large degree, this appears incorrect and instead provides important feedback to the cortex and striatum to be used for further planning (see Grillner et al., 2020).

4.2.3 Striatum, the Input Structure of the Basal Ganglia, Receives Modular Input from Different Parts of the Cortex

The striatum, the input structure, is the main processing center of the basal ganglia, and it is much larger than other compartments of the basal ganglia. In contrast to the spontaneous activity of the SNr/GPi, the SPNs are

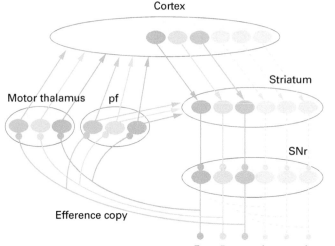

Cortex

Striatum

Motor thalamus   pf

SNr

Efference copy

Eye   Posture   Locomotion.......

Figure 4.7

Schematic representation of the basal ganglia's downstream control of motor structures in the midbrain-brainstem, with specific efference copy information transmitted back via the thalamus to the cortex and striatum. Subpopulations of neurons in the cortex project to subpopulations in the striatum, which in turn inhibit discrete groups of neurons in the SNr (Foster et al., 2021). Each circle indicates groups of neurons. Note that the upstream axonal branches to the motor thalamus and the PF nucleus forward efference copies about the specific activity in the output channels. Only the direct pathway connectivity between the striatum and SNr is included in this scheme. McElvain et al. (2021)'s contribution is that the SNr is subdivided into subpopulations with specific motor targets and that each conveys an efference copy to different parts of the motor thalamus or PF, and further to the cortex and striatum.

hyperpolarized and silent at rest and designed to have a high threshold for activation because they express potassium channels of the inward rectifier type (Kir), which keeps them hyperpolarized at rest. When the SPNs become depolarized by input from the cortex, for instance, the Kir channels will be closed, and then the SPNs become much easier to activate.

The activation of striatal neurons is in turn primarily driven by excitatory glutamatergic input from the cortex and thalamus. Studies on primates and carnivores have shown that different parts of the cortex target different areas within the striatum, but a recent detailed study by Foster et al. (2021) in the mouse has uncovered a degree of precision that was unanticipated and shows very precise modular organization. Figure 4.8a shows that the somatomotor parts of the cortex, representing the primary and secondary motor cortex and the corresponding sensory areas, are subdivided into areas representing the forelimb, trunk, hindlimb, and orofacial areas, and each of them project to correspondingly well defined discrete modules within the DLS (figures 4.8a, 4.8b, and 4.3e) and the ventral parts of the dorsal striatum. Figure 4.8b shows the dorsal striatum with the areas for the trunk (tr), lower limb (ll), upper limb (ul), and large areas for the inner and outer parts of the mouth (mi amd mo).

The direct pathway neurons from the different parts of the DLS target specific parts of the SNr (figure 4.8a) and maintain the same degree of subdivision, as does the indirect pathway that projects with maintained specificity to the GPe, and then to the appropriate compartments in the SNr, which appear somewhat less specific. The DLS is involved specifically in the control of movements and include learned motor sequences (referred to as "habits"). These sequences are stored and become almost hardwired and difficult to modify.

The same type of specific connectivity is maintained for the associative areas of the cortex (shown in mauve and green in figure 4.8a), which project to the DMS and the direct and indirect projections to the medial parts of the SNr and GPe, respectively. The DMS receives input from the prefrontal and limbic areas, and the ventral striatum from the hippocampus and limbic areas. The DMS and ventral striatum are thought to process information of a more cognitive and emotional nature. In primates, goal-directed behavior is controlled by the DMS and the corresponding rostral parts of the basal ganglia (i.e., learned motor sequences that can easily be modified as time passes or the external conditions change, such as if one item is rewarding for a period and then instead becomes punishing).

The collaterals of the SNr project to the PF nucleus in the thalamus with a maintained modular organization, and then back to the same areas of the cortex and striatum (not shown in figure 4.8a), so that the information concerning the SNr commands to downstream motor centers are signaled back to the part of the cortex that "talks" with the corresponding module in the striatum. This shows that the cortex and the various compartments of the basal ganglia are subdivided into an intrinsic modular organization. There is well-organized feedback from the SNr to the ventromedial thalamus, which

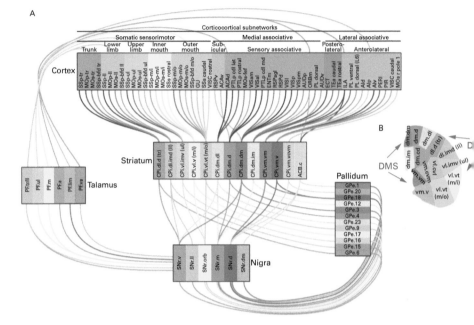

Figure 4.8

Alignment between different parts of the cortico–basal ganglia–thalamic loop. *A*. In the cortex, two medial associative subnetworks of exteroceptive sensory areas, two lateral associative subnetworks of interoceptive limbic areas, and five somatic sensorimotor subnetworks of body regions project into largely distinct striatal subnetworks, whose outputs form the more parallel indirect (striatopallidal) pathway and the more convergent direct (striatonigral) pathway. The pallidal domains send convergent projections to the same nigral domains targeted by their input striatal domains. The nigral domains then project to six regions of the PF thalamus, which in turn are interconnected with the originating cortical regions. *B*. A section through the dorsal striatum showing the compartments within the striatum and color-coded as in *A*. For the DLS, the modules for the trunk (tr), lower limb (ll), upper limb (ul), and two ventrolateral areas for the mouth, the inner (mi) and the outer (mo). Abbreviations: Cortical areas: A, associative cortex; ACA, anterior cingulate area; AI, agranular insular; AUD, auditory; DMS, dorsomedial striatum; DLS, dorsolateral striatum; ECT, ectorhinal; ENT, entorhinal; GU, gustatory; ILA, infralimbic area; MOp, primary motor cortex; MOs, secondary motor cortex; PERI, perirhinal; PIR, piriform; PL, prelimbic; PTLp, posterior parietal association; RSP, retrosplenial; SSp, primary somatosensory cortex; SSs, secondary somatosensory cortex; TEa, temporal association; VIS, visual; VISC, visceral. Striatal areas: ABCc, core of the nucleus accumbens; CP, caudoputamen; CPc, caudal CP; CPi, intermediate CP; CPr, rostral CP. Thalamic areas: PF, parafascicular nucleus; PF.a, associative PF; PF.lim, limbic PF; PF.m, mouth PF; PF.tr/ll, trunk/lower limb PF; PF.ul, upper limb PF; PF.vs, ventrostriatal PF. Nigral areas: SNr.d, dorsal SNr; SNr.dm, dorsomedial SNr; SNr.ll, lower limb SNr; SNr.m, medial SNr; SNr.orb, orobrachial SNr; SNr.v, ventral SNr. Modified from Foster et al. (2021).

is also subdivided into specific modules (not indicated in figure 4.8) and then project back to the sensory and motor areas of the cortex.

The input to the striatum from the nuclei in the thalamus represents around 40 percent of the excitatory input to the striatum (Doig et al., 2010). The PF is the largest nucleus and is subdivided into specific somatomotor, associative, and limbic compartments projecting to the dorsolateral, ventral, and associative (dorsomedial) parts of the striatum, respectively (Mandelbaum et al., 2019), as well as back to the cortex in an orderly fashion. The PF is critical to the performance of learned motor programs (Wolff et al., 2022; Dhawale et al., 2021) and receives input from a variety of subcortical structures and the cerebellum (Grillner et al., 2020).

4.2.4 Synaptic Properties of Cortical and Thalamic Neurons Targeting Striatal Projection Neurons

Starting from the detailed anatomy of the basal ganglia, it is now time to discuss the next level of granularity of cell types and connectivity. The neurons that target the striatum are illustrated in figure 4.9. The efferent pyramidal tract (PT) neurons, also referred to as "corticobulbar" or "corticospinal," originate from

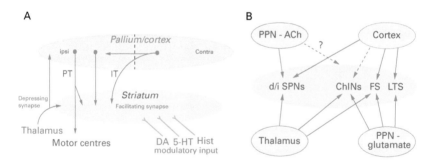

Figure 4.9

Input to neuronal subpopulations in the striatum. *A.* Many cortical/pallial axons that target the brainstem and spinal cord (PT-type) give off collaterals to neurons within the striatum. There is a subset of pyramidal neurons that have intratelencephalic axons projecting to the contralateral, as well as the ipsilateral cortex/pallium (IT-type), which also target the striatum. *B.* Cortical and thalamic neurons target both direct and indirect striatal projection neurons (dSPNs/iSPNs) and the ChINs, FS, and LTS interneurons. The glutamatergic pedunculopontine (PPN) neurons only project to the interneurons, whereas the cholinergic PPN target the dSPNs/iSPNs. The red dashed arrow from the cortex to ChINs indicates a variable and weak effect.

layer 5 in the cortex and target motor structures in the midbrain, brainstem, and spinal cord. Their axons course through the striatum on their way downward and give off collateral synapses on the two types of SPNs and several of the interneurons as well. PT neurons provide a major source for excitation of the striatal neurons. In addition, the corticospinal subpopulation of neurons sends off collaterals to the dorsolateral sensorimotor area of the striatum (Nelson et al., 2021; Filipovic et al., 2019). PT neurons activate both the dSPNs and iSPNs, but the dSPNs get twice as large excitatory synaptic current as iSPNs. Moreover, the corticospinal neurons appear to be subdivided into two populations—one that preferentially targets dSPNs and one that targets iSPNs.

Equally important is another class of neurons in layer 5, the intratelencephalic (IT) neurons, which project bilaterally within the cortex and in addition to the striatum, but do not extend farther downstream. Each IT neuron has a very extensive area of termination within the striatum that includes a significant part of the DLS, while the PT neurons have much smaller but very dense focal areas of termination within the DLS (figure 4.10). Both IT and PT neurons make monosynaptic glutamatergic synapses on the spines of SPNs, but in contrast to the PT neurons, the IT neurons provide equal excitation to both types of SPNs. An activation of IT neurons will thus enhance the excitation in a broad area of the dorsal striatum, while PT neurons will target one or several focal areas with a dense synaptic innervation.

To investigate how PT and IT neurons affect behavior, mice were trained to manipulate a joystick along a path to obtain a reward, while the activity of IT or PT neurons could be recorded or manipulated optogenetically. A blockade of IT neurons led to a disruption of the movement, whereas inactivation of PT neurons had only a modest effect (Park et al., 2022). This is a very important finding, in that the IT neurons that have no downstream projections except to the striatum nevertheless exert the most prominent effect on the performance of the movement, whereas PT-neurons, which also target the downstream motor centers, have less effect on the motor behavior. The behavioral prominence of the effects of IT neurons may relate to the fact that their axonal termination field is much broader than that of PT neurons (figure 4.10), but also that they excite PT neurons at the cortical level, and a blockade of IT will indirectly affect the activity level of the PTs.

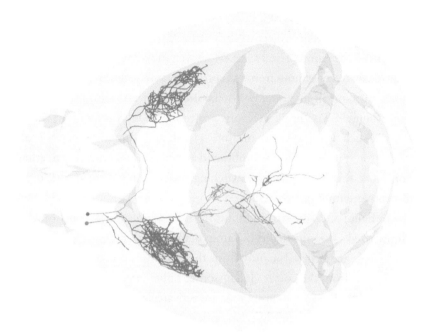

Figure 4.10
Terminal fields of PT and IT neurons in the striatum. The PT neurons are depicted in red and the IT neurons in blue. Note the extensive bilateral terminal arborizations of one IT neuron. Source: Janelia Farm Research Campus, Howard Hughes Medical Institute.

It is interesting to note that the termination area of the thalamic PF neurons is broad, similar to IT neurons, and there are three subtypes projecting to the different parts of the dorsal striatum (DLS, DMS, and limbic part) (Mandelbaum et al., 2019; Ellender et al., 2013; Doig et al., 2010). The synapses are formed on the spines of SPNs, but also on the dendritic shafts, and they activate preferentially N-methyl-D-aspartate (NMDA) receptors. Another interlaminar nucleus, the central lateral (CL) instead terminates in a similar way as do the PT neurons with focused terminations on the spines of the SPNs and with a lower ratio of NMDA to α-amino-3-hydroxy-5-methyl-4-isoxazolepropionic acid (AMPA). CL cells give rise to larger EPSPs and are more efficient in activating SPNs than are PF neurons.

The PF nucleus is by far the largest, and it receives massive input from many sources, such as the cerebellar output nuclei, SNr, in an ordered fashion (see figure 4.7), SC, PPN, brainstem reticular nuclei, and cortex. The PF neurons that project to the cortex (e.g., motor areas) also receive input from the same cortical areas. Moreover, the striatal targets of these PF neurons also receive input from the cortical areas targeted by the PF-cortical axons, forming an intricate, well-structured, cortico-thalamic-striatal network (figure 4.7). The PF neurons projecting to DLS respond to salient stimuli, such as unexpected auditory, visual, and somatosensory stimuli. PF neurons projecting to the limbic and associate areas contribute to attention processing and behavioral switching. Inactivation of these parts of PF disrupts attention processing.

In figure 4.9b, the striatal inputs from the various types of neurons are summarized. The cortex and thalamus activate both types of SPNs, while the ChINs are targeted mainly from the thalamus, but they also receive some input from the cortex (Johansson & Silberberg, 2020; Morgenstern et al., 2022). The LTS interneurons are activated exclusively from the cortex, and the FS interneurons from both structures. An additional input to the striatum is from the PPN, which has two parts (Martinez-Gonzalez et al., 2011). One, using glutamate as the neurotransmitter, targets only the striatal interneurons and is thereby able to provide strong inhibition of the entire striatum through GABAergic action. The other, the cholinergic part of the PPN, projects to SPNs and might act through nicotinic or muscarinic receptors, or through both. They complement the ChINs, and it should be recalled that the cholinergic innervation of striatum is very extensive and nearly as dense as the dopaminergic innervation, and yet the cholinergic cells in the striatum represent only 1 percent of the total number of cells (Burke et al., 2017).

In conclusion, the cortex and thalamus provide the bulk of the excitatory input to SPNs. The SPNs also receive excitation from the cholinergic PPN, while the glutamatergic input of the PPN targets only the interneurons of striatum. Moreover, the thalamic CL neurons are similar to cortical PT neurons, having patchy, focused terminations and larger synaptic effects compared to PF neurons, which resemble cortical IT neurons, with a broad termination field, and provide smaller synaptic effects.

4.3 SYNAPTIC INTERACTION WITHIN THE STRIATUM

To analyze the synaptic interaction within the striatum, as well as the properties of the predominantly GABAergic synapses, the method of simultaneous recordings of pairs or triplets of interacting neurons is preferred, though it is demanding (Planert et al., 2010).

The connectivity ratio between the two types of SPNs, FS and LTS, and the ChINs differs when the neurons are less than 50 μm (red) or 100 μm (blue) apart (figure 4.11a; Planert et al., 2010; Taverna et al., 2008). It is important

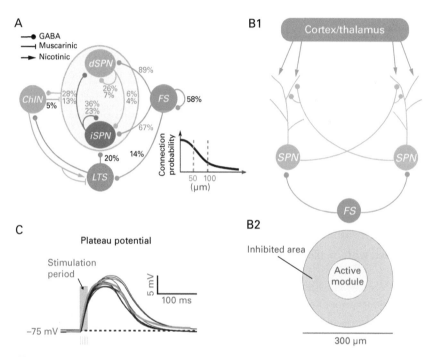

Figure 4.11

Striatal microcircuit. *A.* Connectivity ratio between the two types of SPNs, the fast-spiking interneurons (FS), the low-threshold-spiking interneurons (LTS) and the cholinergic interneurons (ChINs). Connection probabilities within and between neuronal subtypes are shown by respective arrows; numbers in red correspond to connection probabilities for a somatic pair at a distance of 50 μm, while numbers in blue correspond to 100 μm. *B1.* Cortical and thalamic inputs to the distal dendrites of SPNs, while FS neurons target primarily the soma, where action potentials are initiated. *B2.* A circle indicating the area that is inhibited from an active module. *C.* Somatic potential response to spatiotemporal clustered synaptic input, demonstrating the model's ability to trigger NMDA-dependent plateau potentials to reproduce experimental findings. From Hjorth et al. (2020).

to appreciate that the synapses that the SPNs make with other SPNs are made on the distal dendrites. The distal dendrites are an important target since this is where the synaptic input from the cortex and thalamus targets the dendrites of SPNs (figure 4.11b1). This means that the excitation may be shunted through the surround inhibition from the nearby SPNs and affect the soma of the cells to a much lesser extent. A strong synaptic activation of the dendrites may also result in NMDA-activated plateau potentials, which are important in that they provide an enhanced depolarizing drive, but they also are vital to synaptic plasticity (figure 4.11c). Also, the plateau potential would be counteracted by SPN-induced lateral inhibition (Du et al., 2017).

When the SPNs within one discrete module are activated, they will inhibit and target the distal dendrites in the surrounding SPNs in an area extending over 300 μm (figure 4.11b2). This means a form of lateral inhibition and that excitatory synaptic input to SPNs in this area will be counteracted if the SPNs in a discrete module within the striatum is activated—a form of "winner takes all."

In contrast, FS interneurons form synapses on the soma of SPNs and therefore effectively influence the spike generation and represent a form of feed-forward inhibition onto SPNs (figure 4.11b1). The FS interneurons are activated from both the thalamus and cortex, and the very same FS interneuron inhibits both types of SPNs (Planert et al., 2010). The connectivity ratio even within 100 μm is very high, and FS interneurons are electrically coupled with each other (Hjorth et al., 2009; Klaus et al., 2011).

4.3.1 LTS: Regulator of Synaptic Plasticity?

LTS interneurons are GABAergic and releasing nitric oxide, and they are spontaneously active at rest and target the spines of distal dendrites of SPNs and ChINs (Frost Nylén et al., 2021). They receive excitation from the cortex, but disynaptic inhibition from the thalamus via a GABAergic interneuron (Dorst et al., 2020). An interesting finding is that the LTS interneurons also target dopaminergic axons within the striatum and exert presynaptic inhibition via GABA$_B$ receptors (figure 4.12c). During learning, the LTS interneurons reduce their level of activity, which reduce the GABAergic presynaptic inhibition of the dopaminergic axons as well, which thus will enhance dopamine release (Holly et al., 2019) and thereby promote learning. It is interesting to note that there is a complementary effect of the reduced presynaptic inhibition of dopamine axons with the reduced inhibition of SPN dendrites, both of which may facilitate learning. On the other hand, the resting activity or an enhanced LTS discharge would counteract the induction of undue learning—under many conditions, such as initiation of standard motor programs, there is no need for learning, of course. The LTS could thus serve as a regulator of synaptic plasticity in the striatum, regulating the release of dopamine from its axonal terminals.

4.3.2 ChINs: A Major Contributor to Striatal Function through Nicotinic and Muscarinic Receptors

The ChINs are large aspiny neurons (figure 4.12b) representing only 1 percent of the cells in the striatum. Nevertheless, they form a very dense network of

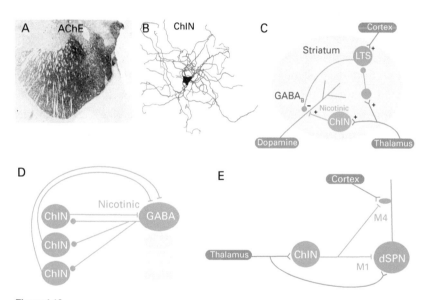

Figure 4.12
The ChINs. *A.* Acethylcholineesterase staining of the rodent striatum. *B.* Reconstruction of an intracellular neurobiotin-filled ChIN with the somata shown in black, dendrites in blue, and the axon in red. *C.* LTS receive direct input from the cortex and indirectly from the thalamus via a GABAergic interneuron. The LTS in turn target dopaminergic axons, which are also targeted by ChINs. *D.* ChINs activate a subtype of GABAergic interneurons via nicotinic receptors. *E.* ChINs activate d/iSPNs via M1 receptors on mainly the somata and M4 receptors specifically on the dendritic spines of dSPNs.

cholinergic terminals in the dorsal striatum (figure 4.12a)—almost as dense as that of the dopamine terminals (Bolam et al., 1984; Mesulam et al., 1984; Wilson et al., 1990; Assous, 2021). They are spontaneously active at a low rate, and therefore they are often referred to as "tonically active interneurons." They are potently activated from the thalamus, and to a variable degree from the cortex. Cortical and thalamic afferents to the striatum express presynaptic nicotinic receptors that may facilitate the synaptic transmission. The ChIN axonal terminals form synapses with the terminals of dopaminergic axons within the striatum that express nicotinic acetylcholine (ACh) receptors (Rice & Cragg, 2004; figure 4.12b). An optogenetic activation of ChINs leads to a release of dopamine via the dopaminergic terminals and represents a local mechanism within the striatum that can regulate the release of dopamine. In this context, it is interesting to note that the LTS are instead inhibited from the thalamus via an interneuron, as discussed previously. The net effect leads to an abolition of LTS-induced presynaptic GABA$_B$ inhibition on dopamine axons, and thus it results in an enhanced dopamine release (figure 4.12b). Considering that the effects from the thalamus, ChINs and LTS complement each other, with one adding presynaptic excitation and the other, when inhibited, reducing the presynaptic inhibition. The net result being that more dopamine is released, which can potentiate learning. In addition to the presynaptic nicotinic receptors, many striatal interneurons express nicotinic receptors, but that does not include SPNs.

Acetylcholine also activates muscarinic G-protein-mediated receptors. They are of two types: the M1 (also M3 and M5), which provide net excitation; and the M2/M4, which are inhibitory (see table 4.1). The M2/M4 receptors provide presynaptic inhibition of corticostriatal and thalamostriatal afferents. An activation of ChINs, either directly or from the thalamic PF nucleus, leads to reduced efficacy of the cortico/thalamic input to SPNs, and thus the presynaptic M2/4 effects dominate over a possible nicotinic effect (Assous, 2021; Ding et al., 2010). M1 receptors are expressed on both types of SPNs and on LTS, which would act in concert with the direct glutamatergic effects that thalamic afferents exert on both types of SPNs.

The ChINs interact in a network, and they tend to be synchronously active at low rates. The underlying network mechanism appears to be that ChINs

Table 4.1

ACh, DA, and GABA receptor expression on striatal neurons and terminals

	dSPN	iSPN	FS	LTS	ChIN	Corticostriatal terminals	Thala-mosttriatal terminals	Dopaminergic terminals
Metabotropic								
M1	+	+		+				
M2/M4	+			+	+^{S}	+	+	+
D1	+		+	+				
D2		+			+	+	+	+
GABAB	+*	+*				+	+	+
Ionotropic								
nAChR			+			+	+	+

*Mainly presynaptic in the striatum (Lacey et al., 2005), GPe, and GPi/SNr.
Abbreviations: S, extrasynaptic. nAChR, nicotinic acetylcholine receptors. + sign indicates the presence of a receptor subtype in a given type of neuron or presynaptic terminal.

activate a subtype of GABAergic interneurons (partially the TH-subtype; figure 4.12d; Dorst et al., 2020) via nicotinic receptors that provides strong inhibition of the local ChIN population. When they become active after the transient inhibition, they tend to fire close to each other, given the background of tonic discharge. The resulting synchronous bursting will be reinforced by the disynaptic inhibitory action. It also means that an excitatory input to all or part of the ChINs leads to a synchronous burst, followed by a pause in the entire network of cholinergic neurons.

The ChINs have a reciprocal relation to the activity of dopamine neurons. When a burst in dopamine neurons occurs as with a salient stimulus or in a reward situation, the ChINs become silent for the duration of the dopamine burst (Raz et al., 1996; Goldberg & Reynolds, 2011). The exact mechanism of the silencing of ChINs is not fully understood, but ChINs have inhibitory D2 receptors that could contribute to this effect.

The silencing of ChINs is most likely important in the context of synaptic plasticity in the following way. The tonic activation of ChINs and M4 receptors on dSPNs under resting conditions leads to a reduced ability of dopamine (via D1 receptors) to enhance the level of cyclic adenosine mono-phosphate (AMP) and cellular effects farther downstream (figure 4.12e; Nair et al., 2015, 2019; Lindroos et al., 2018), being important in the context of synaptic plasticity. By transiently inhibiting the ChINs, and thereby the activation of M4 receptors, the dopamine effect on the D1 receptors will be markedly potentiated, and therefore there is the possibility of an induction of synaptic plasticity in the corticostriatal synapse (Bruce et al., 2019; Reynolds et al., 2022). The actions of M4 and D1 receptors are thus antagonistic at the molecular level in the corticostriatal synaptic transmission. The continuous activation of M4 receptors under resting conditions thus counteracts the induction of synaptic plasticity, which may well be an important function, together with the LTS effects discussed here, in not promoting plasticity and learning when not called for.

In Parkinson's disease, the complex relation between the dopaminergic and cholinergic systems also becomes manifest. With a lack of dopamine, the cholinergic system appears to further enhance Parkinsonian symptoms, and in the pre-DOPA era, muscarinic antagonists such as atropine were used as a medication, but they had unfortunately prominent side effects. The unbalanced cholinergic drive may lead to an enhanced inhibition (M4) of corticostriatal excitation to the striatum, and a lack of inhibition of ChINs during salient or reward situations with a lack of dopamine bursts may also contribute to the symptoms.

In conclusion, ChINs provide a very dense network of axonal terminals within the entire striatum that is almost as dense as that of dopaminergic terminals. ChINs act through both nicotinic and muscarinic receptors and counteract some of the dopaminergic effects on synaptic plasticity. There is a reciprocal relation between the activity of SNc cells and ChINs; when dopamine neurons provide a burst, it leads to a pause in ChIN activity.

4.3.3   The Dopamine System Signals Salient Stimuli, Reward, or Lack of Reward

The dopamine system is ancient and the connectome in figure 4.13 of the afferents and efferents of the SNc/VTA applies surprisingly to both mammals and lampreys, although they represent two phylogenetically extreme

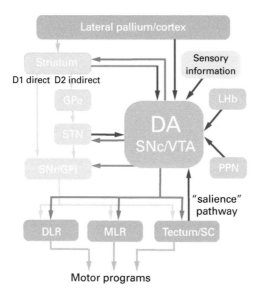

Figure 4.13
The SNc connectome in lampreys and mammals. The efferent and afferent connectivities of the SNc is virtually identical in lampreys and mammals. Thus, the dopaminergic neurons within SNc project to the same structures in the basal ganglia as in mammals and the same midbrain motor centers. The input to SNc is similarly identical from the striatum, STN, cortex/pallium, PPN, and the lateral habenula. From Pérez-Fernández et al. (2014).

groups among vertebrates (Grillner & Robertson, 2016). Dopamine neurons receive input from the SC (salient stimuli), the lateral habenula, the pedunculopontine nucleus, and the cortex/pallium. The most prominent output from the SNc/VTA is to the striatum, but there is also a less dense dopaminergic supply to other structures of the basal ganglia, to downstream motor centers in the SC, and to locomotor command centers (Pérez-Fernández et al., 2014, 2017; Ryczko et al., 2020). The latter aspects mean that the motor centers in the midbrain/brainstem will be primed by the dopamine activity before the motor manifestation of the basal ganglia has had time to become manifested. They will thus increase the excitability in these motor centers before further control from the basal ganglia becomes apparent. The dSPNs, LTS, and FS depend on D1 receptors, while iSPNs and ChINs express D2 receptors, as do the presynaptic terminals of corticocortical and thalamocortical axons and dopaminergic neurons (see tables 4.1 and 4.2).

Table 4.2
Dopamine receptors expressed on neuronal subtypes within the basal ganglia

Striatum	GPe/STN	Substantia nigra
dSPN D1	Prototypic D2	SNr D1
iSPN D2	Arkypallidal ?	SNc D2
FS D1	STN D1/D5	
LTS D1		
ChINs D2		

Note: Question mark after "Arkypallidal" means that the receptor subtype is unknown.

The dopamine system plays a very important role in transmitting the reaction to salient stimuli to the motor apparatus, but also the reaction to an expected or unexpected reward or the converse, the reaction to a predicted reward that did *not* occur (see figure 4.14). Dopamine neurons are tonically active at rest and become further activated by salient stimuli detected in the surroundings or in the reward situation, and conversely become depressed if an expected reward is *not* received.

The SNc particularly targets the sensorimotor part of the dorsal striatum, the DLS in rodents, and the caudal part of the putamen, which corresponds to the DLS in primates. This part of the striatum has an exceptionally high degree of dopamine innervation, and even a single SNc dopamine neuron in humans may support as much as more than a million synapses (Diederich et al., 2020). This finding presumably provides an explanation for why these neurons are particularly vulnerable in Parkinson's disease due to the excessive energy demand on a single neuron.

Salient stimuli activate the SNc, and whenever a mouse engages in a bout of locomotor activity, it is preceded by a burst in the dopamine neurons (da Silva et al., 2018), which enhances the excitability in dSPNs broadly in the DLS through D1 receptors, while iSPNs are inhibited through D2 receptors within the DLS (figure 4.14). This will contribute to the initiation of the motor act, which will be determined by concurrent inputs from the cortex or thalamus that define the specifics of which striatal neurons

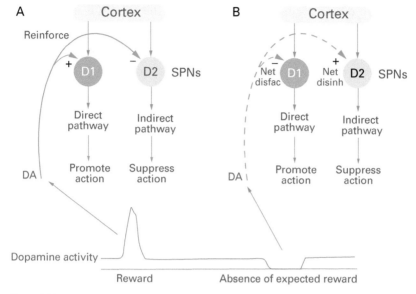

Figure 4.14

The effects of enhanced or decreased dopamine activity on the direct and indirect pathways through the basal ganglia. *A.* Enhanced dopamine activity excites the striatal projection neurons of the direct pathway that express dopamine receptors of the D1 subtype, while it inhibits those of the indirect pathway through their D2 receptors. *B.* Illustrates the opposite situation, with decreased dopamine activity that removes excitation from the direct pathway and reduces the inhibition of the indirect pathway, and thereby it indirectly increases the net excitation of SNr and the resulting inhibition of downstream motor centers.

become activated and the motor act elicited through downstream circuits. An optogenetic activation of dopamine neurons in SNc increases the likelihood of a locomotor burst being initiated. In contrast, if dopamine neurons are activated during an already ongoing movement, there will be no effect on the movement. The timing of the SNc burst is thus important in relation to the initiation of movement. The global dopamine activation of DLS will affect a large part of the striatum, while the specific input from the cortex/thalamus determines which specific module within the striatum is activated, and what movement is produced. Enhanced activity in the ChIN population may also signal the transition to a locomotor state (Howe et al., 2019).

There is also a form of learning that occurs within the DLS that relates to habits (as discussed next); that is, networks producing specific patterns of motor coordination that are stored within the DLS. Whenever they are formed, the network's underlying habits are difficult to change and considered as almost permanent. They may relate to specific patterns of coordination that we do in our daily life when we open the door to the basement, for instance.

The dopamine neurons in the SNc and VTA are not a homogenous group, and there are seven distinct genetically determined subtypes of dopamine neurons in the VTA/SNc that have specific projection patterns (Tiklova et al., 2019) and can be activated under various situations. Some dopamine neurons coexpress glutamate, and others γ-aminobutyric acid (GABA) (Granger et al., 2017; Papathanou et al., 2018; von Twickel et al., 2019), which may complement the action of dopamine. VTA neurons respond in reward situations, but also to the absence of a predicted reward. VTA neurons are clustered into subgroups responding to different sensory or cognitive stimuli in addition to reward (Engelhard et al., 2019). Certain neurons also respond to aversive stimuli. The SNc neurons respond to salient stimuli, but also to rewards. Several types of dopamine neurons located within the VTA innervate the DMS, the ventral striatum, and the frontal lobe. The DMS receives input from the prefrontal lobe and the association areas of the cortex and is engaged in goal-directed behavior (see section 4.4.4.1), in which the behavior is more adaptable, as if a reward occurs during a movement in one direction for a while, and some time later, a movement in a different direction will be rewarded. Finally, the ventral striatum receives input from the limbic system and hippocampus.

4.3.4 The 5-HT System Provides an Evolutionarily Conserved Projection to the Dorsal Striatum and Contributes to Impulse Control

The serotonergic (5-HT) dorsal raphe projects to the striatum, but the innervation is less dense than that of the dopamine system. The 5-HT system is conserved from lampreys to primates (Grillner & Robertson, 2016) and appears to function in conjunction with the dopamine system. The 5-HT neurons are activated at the initiation of behavior and in a reward situation (as dopamine) and tend to remain active throughout the entire behavior, whereas the dopamine neurons tend to elicit short initial bursts before and at the onset of behavior. In a behavioral choice situation, increased activity of 5-HT neurons will lead to a maintained attempt to wait for a potential reward rather than abandoning a test session (Miyazaki et al., 2014; Fonseca et al., 2015; Iigaya et al., 2018). This can be regarded as improved impulse control, a very important quality in everyday life (not least for humans).

4.4 INTEGRATED FUNCTION OF THE BASAL GANGLIA

4.4.1 The GPe and the STN Control Activity in the SNr/GPi

From the modulator systems of dopamine and 5-HT, we will now discuss the roles of two critical comparments of the basal ganglia—the intrinsic nuclei of GPe and STN. The GPe represents the second-largest part of the basal ganglia after the striatum but has only 1.7 percent of the cells of the rat striatum. The STN is even smaller, with only 0.5 percent of the number of striatal cells (Oorschot, 1996). Therefore, the STN that excites the SNr/GPi inhibits movement, and conversely, lesions in the STN can cause involuntary movement. The STN has mainly projection neurons and is divided into three parts: a motor part with input from the cortical motor areas targeting the SNr/GPi and concerned with the control of limb and eye movements, a limbic part, and a cognitive part with input from the limbic and cognitive areas, respectively (DeLong et al., 1985; Steiner et al., 2019; Haynes & Haber, 2013; Grillner et al., 2020). The cognitive part of STN targets the most medial part of the SNr/GPi that is also the target of the DMS and the ventral pallidum. Deep brain stimulation as a treatment of the symptoms of Parkinson's disease often target the STN, and it is then critical that the correct compartment of the STN is stimulated (Bergman et al., 1990, 1994); if not, there can be unfortunate emotional or cognitive side effects of the stimulation (as discussed further later in this chapter).

The neurons within both the STN and GPe are all rhythmically active at rest, which means that they have the advantage that their activity can be modulated in both enhanced and depressed directions. This is important when judging the mode of operation of these nuclei. The most common cell group in GPe is the GABAergic prototypic neurons (expressing Nkx2.1,

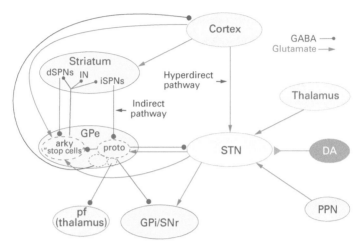

Figure 4.15

Prototypic and arkypallidal GPe neurons. Schematic showing the connectivity of the prototypic and arkypallidal "stop cells" in GPe. Note that the arkypallidal cells project back to all the sub-populations of striatal cells.

Lhx6, and often PV) that are part of the indirect pathway. The connectivity between neurons within the GPe and STN is encapsulated in figure 4.15. When the iSPNs become activated, they inhibit the prototypic cells, which reduce the inhibition exerted on the tonically active neurons within the SNr/GPi (disinhibition), which in turn increases their firing rate and further inhibits downstream motor centers (figure 4.16). Since the prototypic cells also inhibit the excitatory cells in the STN, and as part of the indirect loop, this disinhibition of STN, also results in an enhancement of SNr/GPi activity. Thus, the direct action of the prototypic cells on the SNr/GPi and the indirect action via the STN complement each other (figure 4.16).

What is counterintuitive, however, is that the STN excite the prototypic cells, which would lead to inhibition of the SNr and a promotion of action. However, this may relate to an intricate balance between the prototypic neurons and STN and short-term plasticity in the synapses (Lindahl et al., 2013). The gain in this STN-prototypic connection is reduced by dopamine and is increased in Parkinsonian experimental models, which may be important for the development of Parkinsonian symptoms like tremor (see section 4.5).

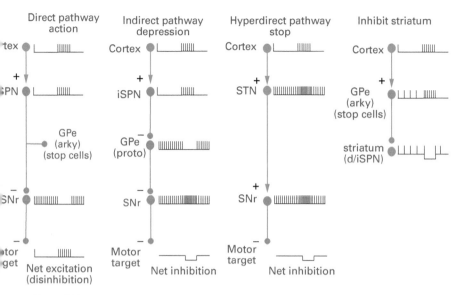

Figure 4.16

The direct, indirect, and hyperdirect pathways. Striatal projection neurons of the direct pathway (dSPNs) directly target the output level (SNr) and enhance the excitability of brainstem motor targets through disinhibition and thus promote action. SPNs of the indirect pathway (iSPNs) will inhibit the spontaneously active GPe, which in turn disinhibit SNr, thus increasing the inhibition of downstream motor targets. The hyperdirect pathway projects to the glutamatergic STN, which in turn targets SNr, which will then inhibit the motor targets. To the right is shown the pathway from the cortex to the stop cells in GPe that projects back broadly to the striatum and act to inhibit striatal circuits.

The second main type of GPe neuron is the arkypallidal neuron (FoxP2), which is spontaneously active at a low rate and inhibited by the proto-typic cells (figures 4.15 and 4.16). When the indirect pathway is activated, GPe neurons will thus become disinhibited. The arkypallidal neurons are GABAergic and project back to the striatum with a massive and extensive network of terminals. A direct projection from the motor cortex provides a very efficient activation of the arkypallidal neurons, and they become activated behaviorally as the mouse terminates an action (Karube et al., 2019; Mallet et al., 2012, 2016; Cui et al., 2021; Abecassis et al., 2020; Ketzef & Silberberg, 2021). They are, therefore, referred to as "stop cells," and they inhibit parts of the striatum very efficiently. When the direct pathway is activated, which promotes action, the stop cells instead become inhibited from the axons of dSPNs, which agrees well with their role to counteract action (Ketzef & Silberberg, 2021).

The STN excites SNr/GPi neurons, and enhanced STN activity will thus cause further inhibition of downstream motor centers. STN neurons receive excitation from both the thalamic PF nucleus and strong activation from the motor cortex (see figure 4.15). The latter is part of the pathway from the cortex via the STN to the SNr/GPi, referred to as the "hyperdirect pathway" (Nambu et al., 1996, 2002). STN neurons also excite the stop cells (arkypallidal neurons; Ketzef & Silberberg, 2021). This action is synergistic with that of the hyperdirect pathway, but on the other hand, the action of STN on the prototypic cells will inhibit SNr neurons (!). The STN neurons and the prototypic cells target the soma of SNr neurons, which means that they directly affect the spike-initiation zone of SNr neurons through direct excitation or disinhibition, and thus emphasizes the priority of inhibition of action. In contrast, the inhibitory synapses of dSPNs that mediate action are located on the distal dendrites of SNr neurons (Smith & Bolam, 1991).

The GPe also contains a separate set of neurons that has input from the dSPNs of the DMS. These neurons have an exclusive target: the PF nucleus of thalamus that is outside the basal ganglia, in contrast to other structures that target the SNr/GPi and STN. The GPe-PF neurons project back to both the striatum and cortex—thereby bypassing the ordinary output route of the basal ganglia (Lilascharoen et al., 2021) and conveying information back to these structures (figure 4.15). The neurons within the GPe-PF pathway are important for conveying information regarding goal-directed behavior and when a rewarded behavior is switched to another goal, as well as the capacity to switch to another action outcome.

In summary, the interactions among the components of the pathways considered here are shown in figure 4.16. The direct pathway is the only one that promotes action, while the other three will in different ways counteract action or balance the action pathway. This emphasizes the importance of controlling the initiation of action and the need for precise control of the termination of action and ascertaining that unplanned movements will not be released through the basal ganglia circuitry.

4.4.2 The Dorsolateral and Dorsomedial Striatum: Separate Roles

The DLS is generally described as the somatomotor part of the striatum with input from somatosensory and motor areas of the neocortex involved in the control of movements. In contrast, the DMS part is more related to goal-directed behavior of the cognitive/limbic nature with input from the visual and association areas of the cortex, sometimes referred to as the "associative branch" of the dorsal striatum. The two areas, although they contain the same general types of SPNs and interneurons, express different molecular markers. In the DMS, Crym AS is found, while the DLS instead expresses GPR155 (Märtin et al., 2019), which may relate to the functional role of DMS and DLS and their input from separate parts of the cortex. The number of neurons in the dorsal striatum in one hemisphere is calculated to be 850,000 in the mouse (1,350,000 in the rat), and somewhat less than half of these belong to the DLS. The dividing line between the population of DLS and DMS is rather distinct in the spatio-molecular map of the striatum.

The different parts of the DLS receive convergent input from the motor and somatosensory areas of the cortex, but separately in a topographic manner from the forelimb, trunk, hindlimb, and orofacial areas (see figure 4.8b), which then become represented in the compartments within the DLS and farther downstream in specific sections of the SNr and GPe. The DLS most likely controls movements that are part of the standard repertoire of innate movements such as turning, reaching, and similar motor patterns. In addition, learned movements referred to as "habits" are controlled from the DLS. These are learned movements adapted to a specific situation, such as opening a special door or learning how to perform a motor pattern in a sport like hitting a serve in tennis. The role of DMS, the associate part of the striatum, relates to goal-directed and more flexible patterns of behavior that can be modified due to changing conditions and depend on input from sensory associate parts of the cortex and the limbic system (see figure 4.8).

4.4.3 Habits and Goal-Directed Behavior in Primates: Striatal Processing

Now, let us consider the integrated action of the basal ganglia in relation to learned motor patterns, in particular what was shown in the very informative

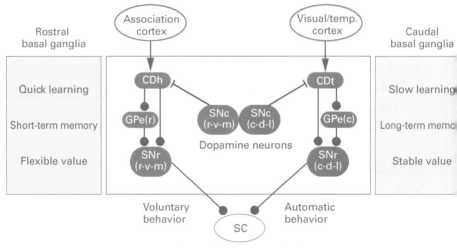

Figure 4.17
Parallel pathways for goal-directed behavior conveyed via the head of the caudate nucleus (CDh) and habitual behavior produced through the tail of the caudate nucleus (CDt). The CDt and CDh receive input from different cortical regions, and both target the SC to elicit saccadic eye movements, although through separate channels and via separate output neurons of the basal ganglia in SNr. Moreover, separate parts of the SNc supply the two circuits. Abbreviations: CDh, head of the caudate nucleus; CDt, tail of the caudate nucleus; c-d-l, caudal-dorsal-lateral; GPe(c), caudal part of GPe; GPe(r), rostral part of GPe; r-v-m, rostral-ventral-medial. Modified and redrawn, with permission, from Kim and Hikosaka (2015).

experiments performed by Kim and Hikosaka (2015). The researchers trained monkeys to associate a few of 100 fractal visual patterns, unknown to them initially, with a reward. They responded by making a saccadic eye movement toward the fractal pattern associated with a reward. When the rewarded pattern came up, dSPNs would fire and downstream SNr neurons would become inhibited, leading to disinhibition of the SC (figure 4.17). Conversely, if a nonrewarded fractal pattern were presented, the indirect pathway would be activated. They showed that the monkeys could remember the rewarded specific fractal patterns over many months. This depended on neurons in the caudal part of the tail of the caudate nucleus, corresponding to the caudal extension of the rodent striatum, caudal to the DLS, which receives input from visual and auditory areas. These caudate neurons project to the lateral part of SNr.

If the monkey were subjected to a situation in which the reward instead changed after a period of time, it had to learn to change strategy and adapt to the new situation. This goal-directed behavior is instead mediated by neurons in the head of the caudate nucleus corresponding to the DMS in rodents (figure 4.17), which project to SNr neurons in the medial part of the SNr. Note that different parts of the caudate nucleus are involved in habitual versus goal-directed behavior, as are different parts of the SNr. Moreover, different sets of dopamine neurons are engaged in the two conditions.

One important contribution of the basal ganglia circuitry is to facilitate the ability to combine individual movements into a sequence in which the various components join each other in a seamless fashion. Parkinsonian patients lose this ability to integrate the movements into one sequence. Instead, they have to perform each movement one at a time. For instance, if asked to pick up an apple on the floor and place it on a shelf in front of her/him, the patient has to bend down, grasp the apple, straighten up again, take a step forward, reach out with the hand, and place the apple on the shelf (Johnels et al., 2001). Thus, only one motor program can be performed at a time, whereas a healthy individual will perform the sequence in an integrated whole. Such a sequence is often referred to as "a chunk" or "chunking" (Graybiel, 1998; Graybiel & Grafton, 2015).

4.4.4 Habits and Goal-Directed Behavior in Rodents: Contribution of DMS and DLS

To unravel the intrinsic function of the striatum, it is critical to record neurons during ongoing behavior. Most studies, whether in primates or rodents, have investigated this activity during specific learned behaviors, followed by rewards, to obtain reproducible conditions. In much fewer studies, the motor pattern is self-induced and remains an open field. The latter type of movements is also important for understanding how standard movements such as locomotion, steering, and orofacial movements are initiated. Many studies, particularly older ones, simply classify the neurons located in the dorsal striatum and active during a certain behavior just as striatal neurons or putative SPNs and FS. More recent studies separate striatal neurons into iSPNs and dSPNs, and the most complete studies also mention whether they are located in the DMS or DLS. Given the recent realization that even neurons in the DLS with input from the motor cortex have subareas for trunk, hindlimb and forelimb, and orofacial areas, a more detailed localization will be important in the future. Next, I will give a few examples from the DLS and DMS.

Jin et al. (2014) recorded neurons in the striatum of mice that learned to rapidly perform five lever presses in sequence to receive a reward. Presumed SPNs could be subdivided into three types that respond when the sequence started, stopped, or had sustained activity during the "chunk" or conversely sustained inhibition. The same subdivision occurred with a similar but not identical distribution, also in the SNr and GPe. They further subdivided SPNs into the two subtypes. The dSPNs could be start/stop subtypes or have sustained activity, whereas the iSPNs were often active at start/stop, but many became inhibited and few had sustained activity. Taken together, it suggests that dSPNs become active during the behavior or in the transition from inactivity to action and back. The iSPNs are also active, particularly at the onset, but many are inhibited during the movement. The results agree with the view that dSPNs promote action and iSPNs inhibit action. The reservation is that we have no knowledge of where in the striatum these neurons were located.

In a subsequent study by Tecuapetla et al. (2016), the activity of dSPNs and iSPNs in the DLS was manipulated optogenetically, while the mice performed a lever-pressing sequence with eight presses before getting the reward. When either type of SPNs were activated or inhibited before the initiation of the first lever press, the initiation was delayed. The more intense the light, the longer the delay would be. This means that an unbalanced and unspecific activation or inhibition of either type of SPNs within the DLS led to a marked delay in the onset of action. Regarding the iSPNs, at high intensity, the initiation of movement would be aborted. The optogenetic interference in DLS, however, will affect a heterogenous group of SPNs, of which some were presumably involved in the normal initiation of the behavior. In contrast, activation of dSPNs during ongoing behavior increased the number of lever pressings, while iSPNs reduced the lever-pressing or abolished the sequence altogether. Inhibition of dSPNs could lead to a slowing of the lever pressing.

Another example is the experiments by the Ölveczsky laboratory (Kawai et al., 2015; Wolff et al., 2022; Dhawale et al., 2021). They trained rats to press a lever twice at a fixed interval of 700 ms to receive a reward, a task that is demanding for them. A rat could learn the task only if its motor cortex was intact. However, if the motor cortex and a large area of the surrounding frontal lobe became lesioned after a task was learned, the rodent could still perform the task perfectly well at the first attempt after the lesion. Clearly, the motor cortex was needed in the learning phase, but not for the execution of the learned motor pattern. If, however, the DLS itself was lesioned, the animal was unable to perform the task. In contrast, if the DMS was lesioned, there was no effect on the performance of the movement. Similarly, if the thalamo-striatal PF nucleus, providing part of the thalamo-striatal input to DLS, was lesioned, the movement could not be performed. Thus, the PF and DLS are required for the execution of this learned motor task under all conditions, but the motor cortex is needed only during learning.

The single lever press is like a reaching movement, and in the learning phase, the rodent starts to perform two lever presses, but that takes too long. During the training phase, the rat gradually develops a movement trajectory with specific kinematics, initially variable but subsequently crystallized to a fixed movement pattern to reach the 700-ms boundary. To investigate if neurons within the DLS contribute to the timing of aspects of the movement, populations of SPNs were recorded during the behavior. It could be shown that different groups of SPNs were active in relation to specific aspects of the movement trajectory that were required to perform the double-lever press within 700 ms. The SPN discharge pattern when learned does not change if the motor cortex is inactivated. This indicates that the learned motor pattern is stored within the DLS and specifies the detailed kinematic structure of the movement.

After the DLS or the thalamic PF nucleus were lesioned, the rat attempted to perform the motor pattern that was previously rewarded but was now only able to generate the initial movement pattern produced before the training with single-lever pressing. Even after intense training, the rat was unable to recover the complex learned motor pattern without either the DLS or the PF. This simple lever pressing resembles reaching and is most likely coordinated by a subset of neurons at the medullary level (Ruder et al., 2021; see also chapter 2 of this book).

There was an early view that when learning a task, the DMS was initially involved and information was later transferred to the DLS, becoming a habit. It is now shown that in the initial learning phase, both the DLS and DMS are involved, but the pattern of SPN activity differs in that more neurons in the DMS become inhibited, while most neurons in the DLS become excited. The pattern of activity in the DLS is then crystalized to neurons that are active at start and stop, during sustained activity during the sequence of lever presses, and to neurons that instead become suppressed. After learning, the pattern of activity in the DLS and DMS remains very similar, but not identical (Vandaele et al., 2019).

4.4.4.1 Goal-directed behavior in rodents

When recording SPNs in the DMS in a push-pull paradigm, the reward occurred in a probabilistic way and could change with the push-pull direction. Nonomura et al. (2018) (see also figure 4.18) reported that different populations of SPNs were activated in the push or pull direction, respectively. Moreover, optogenetically identified dSPNs (electrophysiology) that were active during a pull movement continued firing when the reward was delivered but stopped immediately if no reward appeared. Conversely, iSPNs had some initial activity when the movement started, followed by no activity. When the reward signal appeared, they had low activity, but with no reward, there was instead a marked increase. Thus, dSPNs encoded reward outcomes, whereas iSPNs encoded a nonreward outcome. After a series of mostly no rewards, the rodent is assumed to change strategy from push to pull, or vice versa. The enhanced activity in dSPNs after the reward predicts that the rodent will maintain the same strategy as before, whereas the activity in iSPNs with enhanced activity in the case of a nonreward—signals that it may be time to change strategy. A partial explanation of this asymmetry may be related to the effect of dopamine, which in the reward situation will boost dSPNs through the activation of D1 receptors, whereas the iSPNs in

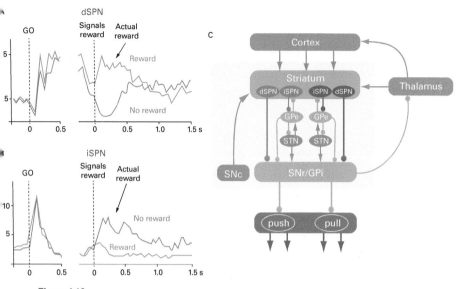

Figure 4.18

The activity of striatal projection neurons of the direct and indirect pathways during a goal-directed push-pull task. *A*. The activity pattern of a direct pathway striatal projection neuron (dSPN) during a push-pull task. The red trace shows a correct response (reward), and the blue trace an incorrect response (no reward). Upon the GO signal, the neuron becomes activated and remains so until a sound signal of whether the response will lead to a reward. The actual reward occurs with a further delay. Note that after the reward signal, the level of activity remains high, while when there is no reward, the activity drops immediately. Redrawn from traces in figure 2F in Nonomura et al. (2018). *B*. The corresponding data for an indirect pathway neuron (iSPN). Note that directly after the GO signal, there is a marked increase of activity that rapidly dies down, while after the no reward signal, there is a marked increase from baseline. Redrawn from traces in figure 2K in Nonomura et al. (2018). *C*. Simplified scheme of the basal ganglia with input from the cortex, thalamus, and SNc to the striatum. The intrinsic nuclei of the indirect pathway are indicated. The synaptic connectivity is indicated in blue for inhibitory neurons and red for excitatory neurons (STN, cortex, and thalamus). Two separate populations of dSPNs and iSPNs control the push and pull motions, respectively. The action is mediated by the basal ganglia output nuclei SNr and GPi to the downstream motor circuits.

the nonreward situation become disinhibited through lack of activation of the inhibitory D2 receptors.

Weglage et al. (2021) reported a somewhat similar experimental situation, but with different results. The mouse had to make a choice in the center port (with a nose-poke) and was then free to choose one of the two side ports with a probabilistic chance to be rewarded (similar to the experiments by Nonomaru et al. 2018). The movements were monitored, as well as the activity in dSPNs, iSPNs from the matrix area, and SPNs from the striosomal area, through Ca^{2+} imaging and expression of GCaMP6. The activity was recorded through head-mounted microscope with the lens within the DMS. They concluded that dSPNs, iSPNs, and SPNs from the striosomal area all share a similarly complete representation of the entire action space, including task- and phase-specific signals of action values and choices. The three types of neurons responded in a very similar way. The fact that the striosomal SPN activity is similar is very surprising, given that they are thought to report the evaluation and control of dopamine neurons. These results contrast with those of Nonomura et al. (2018), which show that dSPNs and iSPNs convey separate types of specific information. The experimental situation is somewhat different (nose-poke versus push-pull actions), and the GCaMP6 recordings with a lens inserted in the brain instead of electrophysiological recordings. Somehow the DMS neurons recorded in the two experimental situations responded in a very different way, for unknown reasons perhaps related to the experimental design.

In conclusion, enhanced activity of the dSPNs within the DLS relates to the initiation and maintenance of the movement, while the iSPNs may be active initially but then drop out. In the DMS, the activity is more complex, presumably related to the goal-directed and flexible nature of the response pattern. In some settings, the dSPNs signal maintained activity after a reward had been received and a maintained strategy, while iSPNs instead increase their activity after the "no reward" signal, indicating a change of strategy.

4.4.5 Value-Based Evaluation of Action through the Control of Dopamine Activity

Mammals, including humans, reach perfection through training by combining microcircuits to develop entirely new motor programs. This includes a progressive improvement, whether we learn to write individual letters or perform whole-body movements like learning to serve in tennis. From a high variability of the initial attempts, we perfect the movements so they become very reproducible. In doing so, we have formed an intrinsic template for the motor coordination, and in this context, the basal ganglia take central importance. This requires that we evaluate the various trials as better or worse, and after a few or many attempts, the motor program has crystalized. The evaluation is mediated to the basal ganglia via the dopamine neurons—when bursting, a reward is signaled that can promote the motor pattern just performed (as discussed next). Reduced dopamine activity instead has the opposite effect. Most likely, the 5-HT system also contributes (Miyazaki et al., 2014). For movement control, the DLS and the SNc take center stage.

The striatum has two compartments (figure 4.3c), as noted earlier. The matrix part deals with the control of action, and the striosomal part is engaged in the control of the level of activity in the dopamine neurons in the SNc and VTA (figure 4.19a). The striosomes are thought to provide a value-based motivational signal related to reward, lack of expected reward, aversive signals, or saliency, and they serve to evaluate how successful a movement has been (Crittenden et al., 2017). It receives input from parts of the pregenual anterior cingulate cortex, caudate orbitofrontal cortex, and the thalamic intralaminar nucleus, but not from the large PF nucleus (Amemori et al., 2021). Microstimulation of these regions induced *pessimistic decision-making* by the monkeys, supporting the idea that the focal activation of these regions induces an anxietylike state that leads to an activation of the striosomes. This is in accordance with the fact that activation of striosomes inhibits the dopamine neurons (figure 4.19). Ontogenetically, the striosomes represent the first part of the striatum to develop, and striosomes are evolutionarily conserved from lampreys to mammals (Stephenson-Jones et al., 2013; Grillner & Robertson, 2016).

The striosomes, as well as many other structures, influence the level of activity in the dopamine neurons (figure 4.19b) as follows:

- The dSPNs in the striosomes inhibit dopamine neurons in the SNc.
- The iSPNs in the striosomes project to a subset of spontaneously active glutamatergic GPi neurons that target the lateral habenula (LHb) and are referred to as "habenula projecting GP neurons" (GPh; Stephenson-Jones et al., 2013, 2016). A subpopulation of LHb neurons projects directly to dopamine neurons in the SNc, but mainly through an

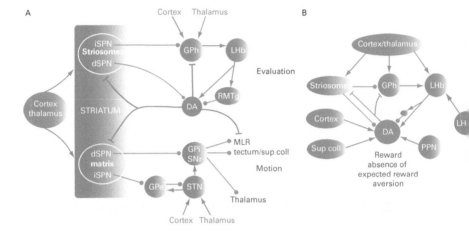

Figure 4.19

Overview of the basal ganglia/habenular circuits underlying the control of motion and evaluation. *A.* The lower motion circuit corresponds to the circuits detailed in figure 4.5. The dSPNs in the matrix compartment target the GPi and SNr, as well as brainstem motor centers, and send a collateral response back to the thalamus (the lower part of the diagram). Also indicated is the indirect pathway via the GPe and STN. The evaluation circuit, in the upper part of the diagram, shows the lateral habenula (LHb), which targets the dopamine (DA) neurons both directly and indirectly via the GABAergic rostromedial tegmental nucleus (RMTg). The LHb receives input from the glutamatergic habenula-projecting globus pallidus (GPh). The GPh receives excitation from the cortex and thalamus, whereas it receives inhibition from iSPNs in the striosome compartment. Dopamine (DA) neurons are inhibited by striosomal dSPNs and send projections to the mesencephalic locomotor region (MLR) and optic tectum/SC (sup coll). The color code is blue for GABAergic, red for glutamatergic, and green for dopaminergic neurons. *B.* Neural mechanisms affecting the DA neurons. Same abbreviations as in A, except for PPN, pedunculopontine nucleus, and LHyp, lateral hypothalamus.

inhibitory relay (RMTg). Another LHb subpopulation target 5-HT neurons in the raphe nucleus, also through RMTg. An increased LHb activity will thus reduce the dopamine and 5-HT activity, but a silencing of the LHb will instead result in an increased level of dopamine activity—that is, a reward (Hu et al., 2020).

• Neurons in the lateral hypothalamus signal aversive stimuli and activate LHb neurons with an inhibitory impact on the dopamine neurons (Lazaridis et al., 2019; Lecca et al., 2017).

- The tectum/SC mediates salient stimuli to the dopamine neurons in SNc (Comoli et al., 2003; Pérez-Fernández et al., 2017).
- The PPN, with input from the SNr and many other structures, target the SNc.
- Dopamine neurons in the SNc receive input from the cortex.

This list tells us that reward, disappointment, and aversive stimuli are being transmitted via the dopamine cells and can originate from many parts of the nervous system. Moreover, if it should affect learning, the signal needs to be transmitted quickly to the neural structures that possibly will be modified through learning.

As discussed previously, the input from the limbic cortex will activate the striosomes that affect dopamine neurons directly, but also the GPh and the LHb and then the dopamine neurons. This is most likely of particular importance for the evaluation process (Stephenson-Jones et al., 2012b, 2013, 2016; Lecca et al., 2017; Lazaridis et al., 2019; Amemori et al., 2021). The striosomes exert direct inhibition of the dopamine neurons and via another part inhibit GPh, which will result in disinhibition of the LHb. The net result in this case is also inhibition of the dopamine neurons. At rest, the GPh, LHb, and dopamine neurons are spontaneously active so that they can be modulated to either increase or decrease the level of activity. The net effect of activation of the striosomes from the limbic cortex would thus be reduced activity of the dopamine neurons, such as occurs in an unsatisfactory movement. However, both the GPh and LHb receive external input from the cortex, and the LHb from the lateral hypothalamus that can further affect the dopamine neurons and even counteract the striosome net effect. The activation of striosomes has been considered as part of a "pessimistic decision-making network in primates" consisting of the frontal cortex and striosomes (Amemori et al., 2021). For a successful life, whether we are talking about rodents or humans, a critical evaluation of the surrounding world may be more important than being overly optimistic.

However, the dopamine neurons also receive a powerful excitatory input from the tectum/SC, through which salient visual or auditory stimuli activate dopamine neurons in the SNc (Pérez-Fernández et al., 2017; Comoli et al., 2003). A study by da Silva et al. (2018) has shown that when a mouse initiates a bout of locomotor activity, it is practically always preceded by a burst of dopamine activity, perhaps initiated by a salient stimulus that has created interest in exploring the origin of the stimulus. Salient stimuli, as well as a reward, can activate neurons in the SNc that innervate the DLS. For the VTA, the reward aspect may be more prominent. It affects the DMS, the ventral striatum, and the prefrontal cortex. This can occur through the excitation of dopamine neurons directly from the cortex, PPN, and other structures or inhibition of the LHb, thereby disinhibiting dopamine neurons. The information conveyed to the dopamine neurons should be value-based and relevant for either a reward or the converse, absence of an expected reward.

4.4.6 Reinforcement Learning and Synaptic Plasticity

A common view is that the basal ganglia contribute to motor learning through reinforcement learning. In the context of the basal ganglia, for instance, a command from the motor cortex can activate a subset of neurons in the striatum that is then further conveyed to downstream motor centers, producing a movement such as a tennis serve. The success is evaluated by circuits in the limbic cortex and pregenual anterior cingulate cortex (as discussed previously) that can control the level of activity in dopamine neurons. If the result was evaluated as successful, the precise neural circuitry used to elicit the movement, such as a direct pathway (see figure 4.19a), is thought to be reinforced through long-term potentiation (LTP) by the dopamine innervation of the matrix compartment. If the coordination is instead judged as unsatisfactory (striosomes), the dopamine activity is inhibited, which in turn may promote a modification of the striatal circuitry employed through long-term depression (LTD). The dopamine signal is sometimes referred to as the "critic," echoing the jargon used in the context of robot control.

The dopamine signal can act on all sites within the striatum in which dopamine receptors are expressed, but it will influence the circuits active at the very moment concerned. One specific site often considered is the corticostriatal synapses of the dSPNs and iSPNs, through which cortex exerts its action. If a cortical command is evaluated as successful, it will be followed after a very short interval by a dopamine burst that will facilitate this synapse (LTP), particularly if paralleled by an inhibition of the resting cholinergic activity by a pause in the ChIN activity. The cholinergic M4 receptor activation counteracts the potentiating effect of dopamine via D1 receptors (Nair et al., 2019); therefore, the concurrent pause in the ChIN activity is important for learning to occur. The dopamine action outlasts the presence of dopamine extracellularly by hundreds of milliseconds due to the intracellular processing initiated by dopamine (Hunger et al., 2020), which is important in the context of synaptic plasticity. The presynaptic depression of synaptic transmission (LTD) in corticostriatal axons depends on an activation of D2 receptors combined with a reduction of the activity in cholinergic M1 receptors (Wang et al., 2006).

The plastic changes of synaptic transmission in the microcircuit concerned occur due to a receptor- and voltage-induced Ca^{2+} entry, together with G-protein-dependent cascades induced by dopamine or other modulators. In the cortico-striatal synapse of SPNs, the Ca^{2+} level increase, leading to LTP that occurs via NMDA receptors or Ca^{2+} permeable AMPA receptors. As important as LTP is the converse process that reduces the synaptic strength, which is LTD that can be induced by endocannabinoids acting presynaptically and also needs postsynaptic activation of L-type calcium channels and metabotropic glutamate receptors (mGluR1s); see Grillner et al. (2020).

In conclusion, reinforcement learning requires a circuit that evaluates whether an action has been successful or not. This can be provided by the limbic and cingulate cortex acting through the striosomes-GPh–lateral habenula to control the level of dopamine activity that can either be facilitated or depressed (see the previous discussion and figure 4.18b). The dopamine system acts through modifying the synaptic strength in the striatal microcircuits that are responsible for the movement (matrix area) evaluated, whether acting through LTD or LTP.

4.5 DYSFUNCTION OF THE BASAL GANGLIA: PARKINSON'S AND HUNTINGTON'S DISEASES AND OTHER CONDITIONS

The importance of the basal ganglia for the overall function of the brain becomes clear when one considers the many neurological and psychiatric disorders that are caused by its dysfunction. The most prominent is Parkinson's disease, which affects 1 percent of the population over the age of 60. Initially, it can be treated pharmacologically with dopamine agonists, but it progressively becomes very debilitating. Many hyperkinesias also involve the basal ganglia and provoke a serious medical condition, such as the devastating Huntington's disease, which is inherited and gives rise to involuntary movements, other dystonias, and hemiballismus; Tourette's syndrome, featuring coordinated integrated motor patterns such as vocalizations and tics that can be triggered in a stereotypic fashion; and obsessive-compulsive disorders, in which an individual may have the compulsion of performing a ritual over and over again, such as whether the oven is on or not (Graybiel & Rauch, 2000). Attention deficit hyperactive disorder (ADHD) is another syndrome of this type; it affects the motor system, attention span, and impulse control. Some psychiatric disorders are also dependent on dysfunction of the basal ganglia, particularly the ventral striatum. Many psychopharmacological drugs act on D2 receptors, affecting iSPNs and the GPe. Recreational use of drugs is linked to the ventral striatum, but compulsive drug taking is thought to depend on the dorsolateral striatum.

In the basal ganglia, only the direct pathway promotes action (which is compromised in Parkinson's disease), whereas the indirect and hyperdirect pathways and the GPe stop cells (see figure 4.16) all counteract movement. These latter pathways are, in one way or another, involved in the generation of hyperkinesia. Here, we will focus on two opposing disorders—the hypokinetic Parkinson's disease and the hyperkinetic Huntington's disease.

4.5.1 Parkinson's Disease

The Parkinsonian patient suffers from symptoms of all parts of the motor system, ranging from postural control to skilled precision movements (Redgrave et al., 2010). Patients tend to bend the body forward and walk with short steps, not lifting the feet as in normal walking (referred to as "a shuffling gait") and have general slowness of movement, called "bradykinesia" (figure 4.20a). The amplitude of the movements is often reduced; remarkably, single cell recordings in the GPi and STN of behaving monkeys performing movements of different amplitudes documented a systematic relation of cell activity to movement amplitude (Georgopoulos et al. 1983). This finding is

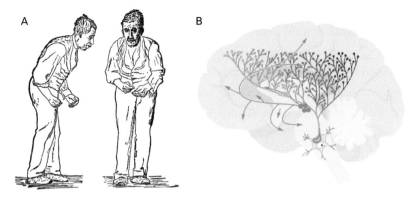

Figure 4.20

The characteristic motor symptoms of Parkinson's disease. *A.* Woodcut representations of front and side views of a man with Parkinson's disease. From Paul de Saint-Leger's 1879 doctoral thesis, as cited and published by Gowers (1886; also see figure 14.5). *B.* Parkinson's disease–prone systems are characterized by hyperbranching axons, regardless of the neurotransmitters. Because hyperbranching neurons project to broad brain areas, the related symptoms are multifaceted. Each dot represents neuronal groups, with projecting axons indicated by arrows. Of note, nonmotor symptoms transmitted by dopaminergic VTA are also included, although disease susceptibility is much higher in SNc than VTA neurons, reflecting the distinctive physiology, bioenergetic control mechanisms, and higher numbers of varicosities in the SNc. From Diederich et al. (2019).

keeping with the prediction that underperformance of these areas would result in reduction of movement amplitude. Saccadic eye movements may have a shortened amplitude, and when a Parkinson's patient is writing, the letters are smaller than usual (micrographia). The movement control may be further affected by tremor in the fingers (pill-rolling tremor) and later in the hands and arms (Gironell et al., 2018), which may even prevent the patient from eating or holding a glass of water without spillage. These patients have difficulty in initiating a new movement, such as walking, recruiting habitual movements, or forming new habits (Redgrave et al., 2010; Schwab et al., 1954). In more severe cases, freezing of movement can occur, in which the patient is unable to continue, such as when walking through a door. Also, the facial expressions connected with various emotions are affected, and the patients may appear to have no emotions, but only the motor representation of the emotions may have been incapacitated. This condition is called "hypomimia," or having a "mask face." As mentioned previously, Parkinson's patients have difficulty with gracefully forming a sequence of motor subprograms into an integrated whole. Thus, when asked to pick up an apple and put it on a shelf in front of him or her, each step (i.e., bend down, reach out to grab the apple, stand up, take a few steps forward, reach out to the shelf, and release the grip of the apple) is performed one at a time, and not as one unbroken motion (Johnels et al., 2001).

The motor symptoms that occur in Parkinson's disorder are due primarily to degeneration of the dopamine neurons that supply the somatomotor part of the striatum (called "putamen" in primates), but several nonmotor symptoms are now also acknowledged, such as cognitive, mood, and sleep disorders (Goetz, 2011; Carlsson, 1964, 2001). A single human dopamine neuron in the SNc that supplies the putamen forms a remarkable number of synapses (varicosities)—over 1 million, which is more than in other mammals (see figure 4.20b; Diederich et al., 2020). This may account for why the dopamine neurons in humans are more vulnerable than those of other vertebrates, given the high energy demand to supply all the metabolic processes required to maintain the function of all parts of the dopamine neurons, mitochondria, signaling pathways, 1 million synapses, and the sodium-potassium pump that maintains the membrane potential. The SNc neurons supplying the motor part of the putamen form a larger number of synapses than those of VTA, supplying the associative and emotional parts of the striatum. This can explain why the somatomotor part of the striatum is the first to suffer in Parkinson's disease. Later, other parts of the striatum are affected, which leads to a variety of nonmotor symptoms in these patients. Except for experimental models, Parkinson's disease is not known to occur in animals other than humans (i.e., in veterinary medicine), and it can be noted that in rodents, the SNc neurons have "only" one-third of the number of varicosities as do humans (Bolam & Pissadaki, 2012) This may explain why the dopamine neurons last throughout the normal life of animals.

The preceding discussion shows how important the basal ganglia (and especially the striatum) are for the normal operation of the motor system. Disorders of the basal ganglia affect all aspects of movement control (i.e., the amplitude of movement, the recruitment of motor programs, coordination between motor programs, as well as motor learning). This illustrates not only that the basal ganglia are of central importance in themselves, but also that the dysfunction of one specific component, the dopamine neurons of the SNc, is sufficient to make the system virtually break down.

Why does dopamine deficiency in the putamen lead to the Parkinson's symptoms? Under normal conditions, an initiation of movement is preceded by a dopamine burst. This will increase the excitability of dSPNs in general via D1 receptors. If at the same time there is a cluster of dSPNs involved in the control of a given type of movement, excited from the cortex/thalamus, the dopamine burst facilitates the recruitment of this specific movement via downstream actions. Without dopamine, as in Parkinson's disease, it will be more difficult to initiate the movement. In addition, the activation of the inhibitory D2 receptors on iSPNs is also lacking, which will enhance the inhibitory action of the indirect pathway at the downstream level via the GPe and SNr. The lack of excitation of dSPNs via D1 receptors and the absence of D2-induced inhibition of iSPNs can account for several of the symptoms of Parkinson's disease with the effects on recruitment of different movements and the reduced amplitude of these movements being recruited. In rodent Parkinson's disease models, the dSPN activity has been showed to be reduced, while the iSPN activity instead is increased, further confirming the reasoning given here (Parker et al., 2018) and explaining why the Parkinsonian brain has more difficulty to initiate a movement with precision.

In addition to the direct effects produced by the lack of dopamine, the dopamine denervation of the dorsal striatum affects the striatal circuitry itself. The distal dendrites of SPNs degenerate, the number of spines decreases, and the dendritic branches are shortened, and thereby they lose synapses that receive input from the cortex and thalamus (Villalba & Smith, 2018). This will further reduce the ability to activate striatal microcircuits (i.e., dSPNs) and thereby reduce the ability to induce movement. A total of 80 percent of the synaptic input to SPNs is located on the distal branch of each dendrite (Hjorth et al., 2020). It is somewhat surprising that the dendrites of the thalamic PF nucleus also are reduced in length. The PF provides a very large part of the thalamic input to SPNs.

Although the focus correctly is on the role of the dopamine system in the pathogenesis of Parkinson's disease, there is also decreased innervation of the striatum from both the noradrenergic locus coeruleus and the serotoninergic raphe nucleus. Although less attention has been given to these two modulatory systems and their role for the striatal circuitry, their decreased innervation of the striatum may well also contribute to the Parkinson's symptoms, although the dopamine innervation of striatum is much denser than either that of 5-HT or noradrenaline.

4.5.2 Pathogenesis of Parkinson's Disease

So far, we have discussed the symptoms and damage to the SNc that occurs with Parkinson's disease. The question now is what pathophysiological mechanism underlies this disorder. There are two mechanisms that may complement each other: one concerned with mitochondrial dysfunction and the other related to accumulation of alpha-synuclein in protein aggregates called "Lewy bodies."

In the autosomal recessive form of hereditary Parkinson's disease, mutations occur in the parkin-gene expressed in dopamine neurons (Mizuno et al., 2001). This gene is important for mitochondrial function. It was later shown that mutations can occur in the mitochondrial DNA of Parkinsonian patients, which may influence mitochondrial function. A conditional knockout of the mitochondrial transcription factor A in dopamine neurons induced a progressive Parkinsonian syndrome in mice (the "MitoPark mouse model"; Ekstrand et al., 2007). González-Rodrígues et al. (2021) developed a conditional knockout of the *Ndufs2* gene belonging to the mitochondrial complex. They could show that symptoms emerge gradually, as in Parkinson's disease, and that the striatal dopamine axons degenerate first, with no release of dopamine in the dorsal striatum (day P30), while the dopamine cell bodies remain intact. This was accompanied by a reduced ability to perform associative learning tasks that are thought to rely on dopamine-dependent synaptic plasticity. Also, fine motor tasks were incapacitated. At this stage, the dopamine cell bodies in the SNc still survive, but a variety of changes occur with expression of ion channels with reduced pace-making and action potential duration, presumably to reduce energy expenditure. Gradually, more extensive symptoms emerge. At p60, the rearing behavior is affected and bradykinesia develops. Only when the dopamine cell bodies degenerate (P100) do gross deficits in gait and other characteristics become apparent. González-Rodrígues et al. (2021) consider the possibility that the soma-dendritic release of dopamine affects the operation of the SNr that expresses D1 receptors (Zhou et al., 2009).

In Parkinson's disease, SNc neurons contain Lewy bodies, which contain a synaptic protein called alpha-synuclein (Spillantini et al., 1997). As proposed by Braak et al. (2003) and shown in several later studies (e.g., Anis et al., 2021), alpha-synuclein aggregations can be transported along the vagal nerve from the enteric mucosa to the brainstem, along unmyelinated axons from the lower brainstem, and then gradually to the raphe nucleus and locus coeruleus, which become affected and give rise to diffuse nonmotor symptoms. This is a process that takes place over several years. Finally, when the alpha-synuclein aggregations reach the level of the SNc, they are taken up by the dopamine neurons; the most vulnerable of the SNc neurons die first and motor symptoms emerge in the form of clinical Parkinson's disease. This unexpected hypothesis has gained support from a variety of experiments (Anis et al., 2021; Rietdijk et al., 2017; Braak et al., 2003), but it is not accepted by all. It should also be remembered that the symptoms of Parkinson's disease can vary significantly, presumably related to which parts of the basal ganglia that are affected by the dopamine denervation.

In conclusion, mitochondrial dysfunction in dopamine neurons can give rise to Parkinson's disease in humans (Parkin) and in experimental models. Since SNc dopamine neurons supplying the putamen have a particularly high metabolic demand, they would be more vulnerable than other dopamine neurons. Alpha-synuclein aggregations occur in dopamine neurons in Parkinson's disease and likely contribute to neuronal degeneration. It is not clear yet if there is a potential link (whether direct or indirect) to the energy supply of dopamine neurons.

4.5.3 Therapy for Parkinson's Disease

With the discovery of dopamine as a transmitter and the fact that a depletion of dopamine gives rise to Parkinsonian symptoms and dopamine was depleted in the SNc in patients with Parkinson's disease, the idea of giving the dopamine precursor L-DOPA to patients was a natural first treatment step (Carlsson, 1964). It turned out that L-DOPA could markedly reduce the symptoms of Parkinson's disease, particularly during the first several years, and ever since, it has been a first-line medication, together with dopamine agonists.

In the early 1970s, when I was lecturing to the medical students on motor control together with my colleague Göran Steg, a professor of neurology at Gothenburg, we used to give a demonstration of a Parkinsonian patient who had volunteered to take part. He would enter the room without having taken his morning L-DOPA; thus he moved slowly and had severe hand tremor, and when asked to drink a glass of water, he was shaking so much that he spilled most of the water. He then took his L-DOPA in front of the audience, left the lecture hall, and came back an hour later. He now walked rapidly and then drank a glass of water with no difficulty or tremor! I believe most of our students remembered the demonstration for many years.

L-DOPA therapy meant a remarkable progress for the patients, but it then turned out that it also had problems. The level of DOPA in the blood of the patient would vary substantially over the day, which also meant that the L-DOPA levels would vary markedly as well, as would the dopamine levels in the striatum. Therefore, the severity of the motor symptoms varied too. They could switch from hypokinesia to hyperkinesia and back again over a few hours. The solution has been to refine the administration of L-DOPA so that similar L-DOPA levels would be maintained throughout the day. This has been achieved to some degree by pharmaceutical means, via slow-release packaging or administering L-DOPA with miniature pumps.

The oscillating L-DOPA/dopamine levels had additional severe side effects, in that the dopamine peaks would lead to L-DOPA-induced dyskinesia (Cenci, 2017). As the dopamine degeneration in the striatum proceeds, the fluctuations increase, and after 4–5 years, it is relatively common that dyskinesias develop, worsening as Parkinson's disease progresses. L-DOPA is transformed into dopamine not only in the few remaining dopamine terminals, but also in 5-HT terminals in the striatum. The dyskinesias may involve facial expressions (e.g., tics) and hand-arm movements and can be very disturbing for the patient (and to some degree for family, friends, and colleagues as well). The origin of the dyskinesia probably relates to synaptic plasticity changes that occur through LTP in the corticostriatal and thalamostriatal synapses. If a motor pattern from the cortex is elicited from a group of corticostriatal axons while the dopamine level is high, synaptic plasticity could be induced, which would potentiate the synaptic transmission from these specific axons. The next time, the threshold for eliciting the motor pattern will have been reduced, and in a final step, a pathologically low threshold has been reached and a dyskinetic movement can be triggered without the intention of the patient. L-DOPA dyskinesias represent a major problem for Parkinsonian patients, but they may also develop L-DOPA-induced nonmotor symptoms that affect emotions and impulse control and cause hallucinations involving other circuits of the basal ganglia, such as the ventral striatum.

Another way to counteract the Parkinsonian symptoms has been to interfere with the subthalamic nucleus (STN). Lesions of the hyperactive STN have been used to counteract the Parkinsonian symptoms by reducing the excitation of the SNr (Bergman et al., 1990). It was therefore counterintuitive that stimulation of STN also would diminish the symptoms of Parkinson's disease (Benabid et al., 2009; Benazzouz et al., 1993). This is now a well-established method that has successfully reduced the symptoms in hundreds of thousands of patients, particularly at a stage when L-DOPA is less effective. High-frequency deep brain stimulation (DBS) is required, and one possible explanation for the effect is that it generates tonic STN activity, thereby abolishing the strong STN oscillations that contribute to the symptoms of Parkinson's disease. Furthermore, the DBS may lead to fatigue in the STN synapses, thereby reducing the excitatory drive onto the SNr. It is critical that the DBS electrode is placed in the small somatomotor part of STN rather than in the cognitive or emotional parts, which can have unfortunate side effects, including personality changes.

Lesions in the motor thalamus have also been used in earlier periods to reduce Parkinsonian symptoms. Such lesions markedly reduced hand-arm oscillations, including tremor, but had little or no effect on other symptoms such as gait control and posture (Duval et al., 2006). This indicates that these movements depend mainly on downstream control via brainstem circuits and do not require thalamo-cortical/striatal feedback, and it is also noteworthy that no cognitive or emotional symptoms were reported by removal of feedback from the output nuclei of the basal ganglia (Duval et al., 2006; Grillner et al., 2013).

In primate and rodent Parkinson's disease models, epidural continuous stimulation of the spinal cord has also been found to relieve the symptoms of experimental Parkinson's disease (Fuentes et al., 2009; Santana et al., 2014). Why would this happen? Epidural stimulation will activate dorsal root afferents, and primarily the dorsal column. This will most likely lead to an unspecific activation of several brainstem nuclei via the dorsal column nuclei and affect the thalamus. The thalamo-striatal nuclei could then be activated to some degree, which would in an unspecific way enhance the overall excitability of the sensorimotor part of the dorsal striatum. Everything else being equal, this could presumably to some degree compensate for the lack of D1 excitation of dSPNs due to denervation of the dopamine neurons, as well as facilitating the initiation of movement. Epidural spinal cord stimulation has been used in the context of other disorders and has recently also been explored in Parkinson's disease patients (Samotus et al., 2018).

4.5.4 Huntington's Disease: A Hyperkinetic Inherited Disorder

Huntington's disease is a devastating congenital and comparatively rare disease (with an incidence of 5 cases out of 100,000 in the United States and Europe), a hyperkinetic progressive disorder caused by a pathologically increased number of cytosine-adenine-guanine (CAG) repeats in the HTT-gene that code for the protein huntingtin. The larger the number of repeats, the earlier the symptoms will occur, usually when the patient is in the late thirites or forties. There is currently no causal therapy for Huntington's disorder, but on the experimental level, different forms of gene therapy are considered that could possibly affect the expression of the HTT-gene (Cheong et al., 2021).

The disease is progressive, and in the early phase, involuntary, sudden, often well-coordinated movements occur. They may include the limbs (chorea), fingers (athetosis), or trunk (hemiballismus). The movements can involve several muscle groups, often in unusual and somewhat bizarre combinations, including facial muscles. In the later phases of the disease, severe mental symptoms and dementia develop and akinesia may occur. The disease is characterized by a progressive degeneration of the striatum, particularly the putamen, but also the cerebral cortex, and a certain reduction in striatal volume can be detected many years before clinical symptoms become manifest (Crittenden & Graybiel, 2011). In addition, neurons in the hypothalamus have recently been shown to degenerate, which can account for some of the nonmotor symptoms that appear (Henningsen et al., 2021).

The initial degeneration in the striatum affects in particular the iSPNs that are at the origin of the indirect pathway, also referred to as the "NoGo" pathway, which results in a reduced ability to suppress movements (Caboche et al., 2017; Parievsky et al., 2017). This can clearly contribute to the motor symptoms that occur in different parts of the body. As the degeneration continues, dSPNs also become affected and akinetic symptoms may occur. Striatal interneurons are reported to be unaffected (Caboche et al., 2017; Parievsky et al., 2017). Huntington's disease in the early phase illustrates what happens if the neural systems that act to stop or suppress movements are damaged. It emphasizes the important role of the nervous system in balancing the precise initiation and termination of a movement episode.

In conclusion, dysfunction of the basal ganglia causes severe problems with initiating and regulating the amplitude of movements and combining movements into an integrated whole, as in Parkinson's disease. Conversely, in hyperkinetic disorders such as Huntington's, dystonia and other conditions, unintended movements of the fingers, hand, arm, trunk, and face (e.g., tics) can be triggered, out of the conscious control of the individual. This illustrates the fundamental role of the basal ganglia in the control of behavior with regard to both initiation and termination of movement.

4.6 THE CONTRIBUTION OF THE BASAL GANGLIA TO THE SELECTION OF ACTION AND THE CONTROL OF MOVEMENT AMPLITUDE

I base my current interpretation of the basal ganglia on older research, and more important, on a number of recent studies demonstrating a detailed organization of the dorsal striatum and SNr (e.g., McElvain et al., 2021;

Foster et al., 2021; Dhawale et al., 2021), which were not available when many of the older views were formulated. The relevant facts are as follows:

- The striatum, the input to the basal ganglia contains 100 times more neurons than the output level SNr/GPi. The intermediate-level GPe has only 1.7 percent of that of the striatum.
- In the somatomotor part of the striatum, there is an orderly arrangement with discrete compartments for the trunk, hindlimbs, forelimbs, outer and inner part of the orofacial region, eye, and orienting movements (Foster et al., 2021; DeLong et al., 1985).
- Each of these DLS compartments project in turn via dSPNs to different compartments within the much smaller SNr/GPi. The SNr in turn contain subpopulations of neurons that project to 42 discrete populations of neurons in the motor centers of the midbrain and brainstem (McElvain et al., 2021).
- SNr/GPi neurons are spontaneously active at rest and inhibit the various downstream motor centers. An activation of a subset of dSPNs will inhibit a group of SNr neurons, and thereby release the specific motor center from SNr inhibition.
- The iSPNs that are localized in the same microregion as the dSPNs project to the prototypic cells in the GPe, which in turn project to the same subpopulation of SNr neurons that are the target of the local dSPNs. The net effect of the iSPNs is an inhibition of the spontaneously active prototypic neurons in the GPe, resulting in a disinhibition of SNr neurons, the converse effect of an activation of dSPNs.
- All downstream-projecting SNr neurons also have a collateral to the PF nucleus in the thalamus that forward information to the matrix portion of the striatum, the cortex, and other thalamic nuclei.
- Each of the SNr populations provides efference copy information that is propagated with maintained specificity to the striatum and cortex. The striatum/cortex is thus informed about the downstream commands issued from the SNr.
- The PPN is also the target of the SNr collaterals, but in this case, they converge without specificity on PPN neurons. This means that it signals

the overall activity in the SNr. The larger the number of SNr neurons that are inhibited when action is initiated, the more PPN will become activated through disinhibition. This means that there is an enhanced PPN-induced excitation of striatal interneurons (as previously discussed), most of which are inhibitory, that provide negative feedback to striatal SPNs.

- Each bout of locomotor activity is preceded by a dopamine burst that will excite dSPNs and inhibits iSPNs.
- Dopamine neurons in the SNc are spontaneously active at rest, activated by salient stimuli, and modulated by reward and by absence of reward.
- Dopamine denervation as in Parkinson's disease leads to the following:
  - Reduced amplitude of movement, such as saccadic eye movements or handwriting (micrographia), a learned movement
  - Difficulty to initiate movement, whether whole-body-movements such as locomotion or fine, precise movements
  - Difficulty in gracefully forming a sequence of motor subprograms into an integrated whole, such as when grabbing a ball thrown at you or just picking up an apple from the ground
  - Difficulty in learning new movement or habits
- Degeneration of iSPNs, as in Huntington's disorder, leads to hyperkinesia, an inability to control involuntary movements of the fingers, hands, arms, and trunk, and facial expressions (e.g., tics) since the indirect pathway is inactivated.
- An inactivation/lesion of the somatomotor part of the STN leads to hyperkinesia due to a reduced excitation of the SNr.
- A learned movement such as a double-lever pressing, with an interval near 700 ms, leading to a reward, requires the motor cortex during the learning period, while when learned, the task can be performed after lesions of the motor cortex and surrounding cortical areas. However, the DLS and thalamic input from PF are always required. Lesions of either structure lead to the inability to perform the task, although a rat can still perform single-lever presses. Lesions of the DMS do not affect the performance of this task.

The conclusions to be drawn are as follows:

- The DLS (caudal putamen) contains a detailed sensorimotor map for the control of the trunk (rearing, posture), hindlimb, forelimb, orofacial area, and oculomotor and orienting movements. This motor map is aligned with the SNr motor map and the downstream motor circuitry.
- The DLS is important for the actual performance of learned motor patterns, whereas the motor cortex is required only during the learning period. These learned motor patterns can be very stable, often referred to as "habits."
- The DMS receives input from the limbic and associate parts of the cortex and has a more adaptive response pattern, as invoked in probabilistic tasks when a reward is given only in a proportion of the test. For example, the condition when a reward is given can change periodically, and a push movement may be rewarded for a period of time and subsequently a pull movement is rewarded instead.
- Whole-body movements such as locomotion and escape reactions also can be elicited from the DMS.
- The dorsal striatum is concerned with the control of both the initiation of action and the amplitude of movement. Studies in behaving monkeys have documented orderly relations between cell activity and the amplitude of arm movements in GPe, GPi, and STN (Georgopoulos et al., 1983).

4.6.1 How Does the Striatum Contribute to the Selection of Actions and the Amplitude of Movements?

Take, for instance, saccadic eye movements in primates. They are triggered by visual input to dSPNs in the caudate tail, which inhibits neurons in the lateral part of the SNr and in turn disinhibits the lateral part of the SC, from which the saccadic eye movements are triggered (Hikosaka & Wurtz, 1983).

Given the detailed sensorimotor input to the compartmental organization of the rodent DLS, it seems likely that subsets of dSPN neurons activated in the forelimb or trunk area of the DLS will inhibit the corresponding SNr compartments. These in turn are able to elicit forelimb or rearing movements of the trunk by utilizing the midbrain and brainstem microcircuits discussed previously. Here, iSPNs are often initially coactivated with dSPNs, but their activity will fade away but be able to terminate a movement. For instance, probably the various DLS motor areas, such as the forelimb part, are further subdivided into reach and grasp areas corresponding to the medullary areas that control reach and grasp.

If the DLS forelimb area would receive excitation from the cortex/thalamus, perhaps due to a salient stimulus, the dSPNs also receive additional excitation via D1 receptors due to the dopamine burst that accompanies a salient stimulus. This could be sufficient to further activate the dSPNs and potentially elicit a downstream motor act.

What about amplitude control? A stronger activation of a set of dSPNs activating a given movement, everything else being equal, will elicit stronger inhibition of the SNr and larger-amplitude movement. Sometimes selection of action and amplitude control have been considered as two alternative roles of the basal ganglia, but this illustrates that the basal ganglia can serve both functions, and it can be noted that in Parkinson's disease, both amplitude and initiation of movement are compromised, as already discussed.

What about selection between stimuli that could elicit a movement? If there are two competing stimuli, both of which could elicit a movement through two different sets of dSPNs, what is the mechanism used by striatal circuits to select one over the other? There is a form of surround inhibition via local axon collaterals between SPNs that target the distal dendrites of the SPNs that are where the input synapses from the cortex and thalamus are located (figure 4.11b). The axonal arbors of SPNs extend over 250 μm in all directions around each SPN. This will mean that if one set of SPNs is activated before another set with input from other sources in the cortex, the latter will already be under inhibitory barrage on their dendrites and the excitation received will be shunted and have less impact. This would mean that the first set will be released, while the second would be suppressed.

This form of surround inhibition would contribute to the selection of action within a diameter of around 0.7 mm. The trunk, hindlimb, and orofacial areas are of similar size (Foster et al., 2021), which means that the axonal arbors of a group of activated dSPNs would be able to provide surround inhibition of a large part of an area such as the striatal forelimb compartment.

If two concurrent stimuli occur within the same area, it would promote the one that is activated the most strongly—a kind of "winner take all."

This form of local surround inhibition can thus contribute to the selection of a motor action within one striatal module within the DLS, such as the forelimb or hindlimb area, but not between other areas of the striatum. One needs, however, to consider that many other mechanisms are at play within the striatum and can affect SPN excitability and whether an action is selected. For instance, the thalamic input can provide direct excitation to selected SPNs, and also mediate synaptic effects via cholinergic and GABAergic interneurons in the striatum. The modulator systems (dopamine, 5-HT and histamine) and the actions from the cholinergic and glutamatergic inputs from the PPN can also profoundly affect the excitability of SPNs and determine whether dSPNs will generate action potentials and thus determine whether downstream centers will be affected. These mechanisms will clearly contribute to the selection process, but as yet they have not been analyzed in sufficient detail.

4.6.2 Is Disinhibition of the Brainstem Targets of the SNr/GPi Sufficient to Elicit Discrete Movements, or Is Complementary Excitation Needed from Other Structures, Such as the Motor Cortex?

The purpose of the tonic inhibition from SNr/GPi of all the midbrain–brainstem motor centers under resting conditions is most likely to keep all the motor centers under control so that they do not become active by accident. This happens, for instance, in the various hyperkinesias, as in Huntington's disorder, when the indirect pathway is not in operation.

As has been noted here, the SNr/GPi represents only 1 percent of the neurons in the striatum, this means that while the dorsal striatum in the mouse has 850,000 neurons, the SNr would be expected to have only 8,500 neurons, given the same proportions as in the rat. Given that there are 42 brainstem motor targets, it means around 200 neurons per target if each target had its own group of SNr neurons. However, some SNr neurons collateralize to several targets. McElvain et al. (2021) subdivide SNr into eight major subgroups, but more subgroups most likely exist, although each target may not have its own SNr subgroup.

Given that a limited number of SNr neurons are available for one downstream target, it may mean that the control may not be sufficiently specific, at least for precise movements. If one considers saccadic eye movements, they can be directed with great precision in many directions, and it seems likely that the SNr neurons assigned to this type of control cannot be as precise as required for each potential saccade. This would open up the possibility that SNr control can be of the permissive type, at least for some types of movement. The SNr neurons projecting to the lateral part of the SC could each cover a limited area of the collicular region controlling eye movements, and excitation from retinal afferents or the frontal eye field (FEF) in the cortex could provide the detailed control of the specifics of the saccade.

One may also ask why the ratio between the striatum and the SNr/GPi is the remarkable 100:1. One possibility would be that striatal cells involved with one specific SNr target could be heterogenous and contain several striatal subpopulations with different sensorimotor inputs requiring different sensory processing, but in each case, they are channeled through the same SNr output stage.

4.6.3 Striatal Control of Innate and Learned Movements

Much of the movement repertoire is innate, as discussed in earlier chapters, such as saccadic eye movements or whole-body movements such as locomotion, but they nevertheless have adapted over time to the development of body shape and weight as the individual grows (Grillner & Wallén, 2004). As we have discussed in chapters 2 and 3, many learned movements are formed by recombining innate microcircuits at the brainstem/midbrain level. The basal ganglia contribute to the learning by providing the combination of tasks. A simple example of a learned behavior is a rat pressing a lever arm twice at a fixed time interval. The execution of pressing the lever arm most likely corresponds to a reaching movement being of an innate nature, but the learning corresponds to the precise timing. The learned motor program is stored in the DLS, and the motor cortex is required only in the initial learning phase. When the memory has been consolidated, it depends on the DLS, and the motor cortex can be inactivated without having an effect on the rat's ability to perform the task (Dhawale et al., 2021).

The process of learning within the basal ganglia is presumably through the general format of reinforcement learning (see the discussion in section 4.4.6). The dopamine neurons signal whether the movement has been a success (i.e., receiving a reward) or not. A movement being trained must generally be repeated many times before achieving perfection. Consider, for instance, a child when learning to form individual letters the first year in school or a new tennis player trying to perform a serve as opposed to an experienced player. During the learning process, a modification is introduced between each trial (Dhawale et al., 2021) that leads to a dopamine reward if the performance was better than in the previous trial, or a decreased dopamine level if the performance was worse. If the reward is low between each trial, the imposed variability is high, but if the reward is high between trials, the variability is kept low. In this way, LTP or LTD can be elicited within the striatal subnetwork engaged in the task, including input from cortico-striatal and thalamo-striatal axons. The superimposed changes in variability make for faster learning than being without the imposed changes, and this applies to both the biological system and to reinforcement learning in technical applications.

For the rat trained to do a double-lever press at a fixed interval (Dhawale et al., 2021), we know that the DLS is required, and from Foster et al. (2021), we know that there is a forelimb compartment in the DLS in which the processing takes place. The forelimb area in the mouse has the approximate size of a cube with a side of 0.7 mm, which would correspond to around 32,000 neurons altogether. This area thus contains around 16,000 dSPN neurons that project to the central lateral aspect of the SNr, where the neurons concerned with the forelimbs are located. In proportion (i.e., 1 percent), this should correspond to roughly only 320 neurons (McElvain et al., 2021)!

4.7 THE ORGANIZATION OF THE BASAL GANGLIA IS CONSERVED FROM LAMPREYS TO PRIMATES

We have discussed the evolutionary aspect on the brain separately (in chapter 1), but here, we consider only some aspects of this topic by comparing the basal ganglia of lampreys with that of mammals. To our surprise, we established that the lamprey has all the organizational features of the mammalian basal ganglia

(Stephenson-Jones et al., 2011; Grillner & Robertson, 2016). There is thus a striatum with input from the cortex (pallium) and thalamus, a subdivision in the matrix and striosome compartments, and a conserved dopamine system. There are a GPe, STN, SNr, and GPi, and they have similar downstream targets and backprojections to the thalamus in the lamprey as in mammals.

Details of the similarities are listed in table 4.3, regarding cell types, transmitters, neuropeptides, and connectivity. In the lamprey, the input to the striatum from the cortex/pallium has both PT- and IT-like cell types, and the striatum contains all the elements of the direct pathway, involving dSPNs projecting to the SNr/GPi. All the components of the indirect pathway, with iSPNs expressing D2 receptors, are also present. Regarding the striatal

Table 4.3

Comparison of the lamprey and mammalian basal ganglia

	Lamprey	Mammals		Lamprey	Mammals
Input to striatum			**Output pathway SNr/GPi**		
cortical IT-type	+	+	spontaneous activity	+	+
cortical PT-type	+	+	direct input D1R/SP SPN	+	+
thalamus	+	+	GABA	+	+
dopamine	+	+	Parvalbumin	+	+
5-HT	+	+	**Indirect loop**		
histamine	+	+	GPe	+	+
Striatal spiny projection neurons (SPN)			direct input from		
D1R/SP	+	+	D2R/Enk SPN	+	+
D2R/Enk	+	+	GABA	+	+
spiny dendrites	+	+	subthalamic nucleus	+	+
Kir	+	+	glutamate	+	+
GABA	+	+	spontanous activity	+	+
DARPP32	+	+	I_h	+	+
rest hyperpol.	+	+			
Striatal interneurons					
cholinergic	+	+			
fast spiking (FS)	+	+			
subtypes of INs	?	+			

Note: INs, interneurons; rest hyperpol., hyperpolerized at rest interneurons, cholinergic and FS interneurons have been identified in the lamprey, but whether some of the newly identified mammalian interneurons exist is still unknown. The arkypallidal stop cells in the mammalian GPe have not been identified in the lamprey.

Furthermore, the lamprey SNc has the same input and output structures as in rodents and the same type of dopamine receptors (see figure 4.13). Dopamine denervation causes similar hypokinetic symptoms in both types of animal (Thompson et al., 2008). The SNc neurons are also activated by salient visual (or other) stimuli (Pérez-Fernández et al., 2017).

Table 4.3 compares the properties of different components of the basal ganglia in lampreys and mammals. Except for cholinergic and FS inter-neurons, it is unclear if the other subtypes of interneurons recently demon-strated in mammals are present in lampreys. The plus sign indicates the presence of a function.

As in mammals, the striatum is subdivided into a number of discrete modules (figure 4.21), connected to specific compartments of the SNr that in turn control individual motor programs (e.g., eye movement; see figure 2.21). During vertebrate evolution, the number of such modules has increased with a gradually more varied movement repertoire, but the basic design of each module has remained virtually the same.

Although the detailed organization of the lamprey's basal ganglia is similar to that of mammals, a major difference is that the number of neurons is orders of magnitude less in lampreys than in mammals. Moreover, we do not know if the striatum is subdivided into a dorsal and a ventral striatum and if the dopamine neurons can be subdivided into an SNc and a VTA subpopulation, although we have labeled them as a putative SNc. However, at the output level, there is both an SNr and a GPi, and they provide tonic inhibition to down-stream targets. It thus appears that all basic features of the basal ganglia were invented early in vertebrate evolution before the evolutionary lines leading to mammals became separate from the lamprey line some 500 million years ago.

In conclusion, we note that the conserved structure and function of the basal ganglia suggest that this structure is essential for the control of behav-ior and the design of this control system has proved so efficient that it has been maintained throughout vertebrate phylogeny (i.e., some 500 million

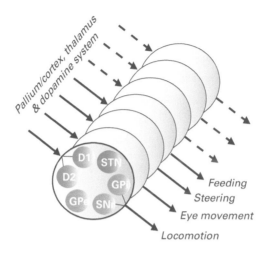

Figure 4.21

Conceptual scheme of a modular organization of the basal ganglia, with one module for each type of motor program. Each module would contain the D1R and D2R projection neurons and the components of the direct and indirect pathway GPi (includes SNr), GPe, and STN. Each module would be activated if sufficient drive occurs from neurons in the pallium/cortex and thalamus. The responsiveness of the modules would be determined by the tonic dopaminergic drive. Whereas the lamprey has a limited behavioral repertoire, mammals (and particularly primates) have a varied and versatile motor repertoire.

years). The basal ganglia are involved in almost all aspects of movement control and the selection of which specific motor program will be active at any given point of time. The striatum is organized in discrete modules for different parts of the body, each of which affects specific modules within the SNr (output of the basal ganglia) that target different midbrain–brainstem motor centers and can promote action through disinhibition (permissive). This is an intricate neural machinery that provides both control of action (matrix) and evaluation of the success of individual movements (striosomes), with the latter being an important prerequisite for motor learning, which represents a critical aspect of basal ganglia function.

The basal ganglia function in close interaction with the parts of the cortex, and they represent two mutually dependent parts of the forebrain machinery. Together they control the motor infrastructure of the microcircuits in the midbrain, brainstem, and spinal cord (discussed in chapter 2). We will learn more about the parts of the cortex in chapter 5.

5 THE ROLE OF THE CORTEX IN THE CONTROL OF MOVEMENT

5.1 INTRODUCTION

The most rostral part of the vertebrate nervous system, the cortex/pallium, contains a motor, a visual, and a somatosensory area in lampreys as well as mammals (Ocana et al., 2015; Suryanarayana et al., 2020, 2021a). This strongly suggests that there is a common basic neural organization in all vertebrates, but the number of neurons in each part of the cortex/pallium differs by orders of magnitude, which also applies to mice and humans. During evolution, however, new circuits have been needed, with the addition of limbs from lampreys to jawed vertebrates, and from mice to primates.

The appendages were mostly used for steering or positioning, as in water-living animals, but on land they became used for locomotion. The interaction with the environment was mostly through the jaws, whether in foraging or fighting. Further elaboration of the visuomotor coordination was required when the forelimbs became used for reaching and grasping (birds such as the osprey use their hindlimbs), an ability that is particularly developed in primates, where delicate control of the hand has evolved along with independent control of the fingers. This has allowed primates (including humans) to finely manipulate objects in the vicinity (whether food or tools), and ultimately in humans for creating arts and crafts.

One important role for the cerebral cortex is to process information about the surrounding world through the primary and associate sensory areas for vision, hearing, and somatosensory information with a retinotopic, tonotopic, or somatosensory representation of the body. These areas are important for the conscious perception of surrounding events. A bilateral lesion of V1 results in blindness in nonhuman primates as well as humans. The cortical processing of sensory information is important for movement control because our movements, whether locomotion or reaching, always relate to the surrounding world. The somatosensory representation is very extensive for the hand and oral area, while it is comparatively small for the trunk and lower limbs. The human hand is a refined sensorimotor organ that can detect subtle changes in the quality of the fine structure of the objects that it is exploring. It is a unique, with a remarkable versatility that allows humans to manipulate objects and learn to play the piano or the violin, for instance, by generating rapid sequences of precise finger movements. In school, humans learn to write letters of the alphabet, whether in Chinese or Swedish, which is a complex task for a child.

The motor representation in the human motor cortex, shown in the well-known illustration by Penfield and Rasmussen (1950; see figure 5.1), was based on data from awake patients undergoing surgery in which the cortex was stimulated. This emphasizes the priority given to the representation of the hands and fingers. The representation of a single finger appears to be larger than that of the trunk and leg combined. Similarly, the representation of the mouth area (including the tongue) is very large. These two regions include the areas of fine motor coordination. The latter is required for speech, with rapid transition in shape of the oral cavity, tongue, and lips as each phoneme is expressed in a dynamic sequence. The adjacent sensory representation of the body in S1 has a similar distorted distribution of the body surface.

The ability to communicate by discrete, mostly innate sounds is widespread among vertebrates. Many primates have developed a more elaborate set of sounds, but only humans are born with the ability to rapidly learn a language spoken in the surrounding environment, or even sign language. Speech requires learning both the syntax of the language and vocabulary, and finally be able to pronounce words accurately and in the appropriate sequence. Specific areas of the cortex are assigned to the interpretation of spoken words (Wernicke's area) and the motor aspect of language (Broca's area), but other parts of the brain are also involved. To utter words requires

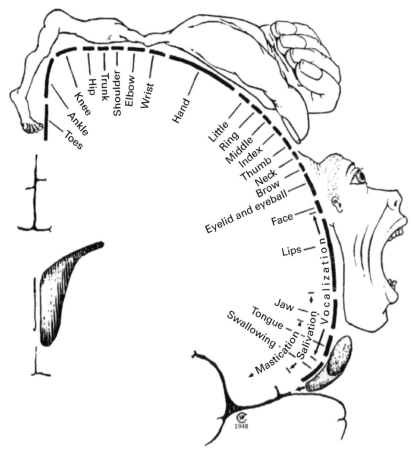

Figure 5.1

Representation of the human M1. Note the large representation of the hand, oral cavity, and tongue—representing fine motor control. Areas for the trunk. lower leg, and forehead, which do not involve any fine control, are represented with much smaller areas. Adapted from Penfield and Rasmussen (1950).

precise control of the flow of air and the shape of the oral cavity and dynamic control of the position of the tongue and lips.

In this discussion, our focus will be on the sensorimotor representations in the cortex, how they interact, their intrinsic function and role, and how they are integrated with other parts of the forebrain and downstream centers for precise control of the motor repertoire. In the context of this book, however, we will not be considering the fascinating areas of perception and consciousness, whether in primates or other vertebrates.

5.2 SOMATOSENSORY AND VISUOMOTOR COORDINATION CRITICAL IN THE PREPARATORY PHASE AND THE TRANSITION BETWEEN DIFFERENT COMMANDS IN A MOTOR SEQUENCE

5.2.1 Vision, together with Cutaneous Sensation, Are Required for Precise Hand Control

To grasp a glass of water in front of you appears to be a simple task, but it requires a series of commands that involve different parts of the brain (Johansson & Flanagan, 2009). One needs to reach for the glass, orient and preshape the hand to the actual form of the glass, smoothly let each finger make contact with the glass, and contract the finger muscles around the glass, lift it without slipping (figure 5.2), and then move the glass to the mouth. The initial reach and hand orientation require visuomotor coordination and a prediction of the approximate weight of the glass. The next phase, the grasping of the object, is to a large degree a sensorimotor event. The tips of all the fingers have

Figure 5.2
The human hand grasping a glass or holding a raspberry. The fingertips have a very high density of slow-adapting (SA; Merkel) and fast-adapting (FA; Meissner) receptors, and the palm of the hand has another set of mechanoreceptors (Pacini and Ruffini). Maintaining a hold on a glass involves both the fingertips and the palm, while for holding a delicate structure like a raspberry, only the fingertips are involved.

an exceptionally high density of fast- and slow-adapting receptors, which is required for registering even minimal contact with an object (Meissner receptors) and for recording the pressure exerted by the fingertips when holding an object (Merkel receptors). The skin of the inner hand has another set of receptors, the fast vibration–sensitive Pacini receptors and the slow-adapting Ruffini receptors (figure 5.2). The receptors on the fingertips are critical and provide information on how firm the grip is, and the receptors within the hand are important for securing the hand's grasp around the glass. If the prediction of weight is correct and the sensory feedback appropriate, the glass can be brought to the mouth. This requires processing in the parietal lobe, the somatosensory and motor cortex, and downstream centers in the brainstem and spinal cord. Even more delicate is the regulation of the force of the finger grip—if you try to grasp a structure like a raspberry and exert too much force, you will crush it, but too little and you will drop it.

If the intention is to lift an object from point A to point B, but you have to circumvent an obstacle while doing so, visual information and the control of gaze become important. The eyes are directed initially to point A, and as the lift starts, the gaze is redirected to the obstacle to ascertain that the trajectory of the hand will avoid a collision. Even before the hand reaches the position of the obstacle, the gaze is again redirected to the final location. In this task, the trajectory is determined by using the anticipatory control mediated through the control of the gaze (Johansson & Flanagan, 2009).

5.2.2 The Sensory Representation in Somatosensory and Other Areas Are in Dynamic Equilibrium

The sensory representation from the skin, such as the receptors in the hand, is transferred via ascending pathways and the thalamus to the somatosensory representation in the cortex (S1). Parts of the body are represented in a map-like way in S1, located just caudal to M1 in the parietal lobe. The hand and fingers have an extensive representation (figure 5.1), whereas the adjacent forearm has a much smaller representation, but the areas are in dynamic equilibrium. If the forearm was anesthetized by applying a local anesthetic cream, the two-point discrimination on the fingertips increased a good deal (Björkman et al., 2009). When monitoring the sensory area in S1 with brain imaging via functional magnetic resonance imaging (fMRI), the sensory hand area had also increased significantly.

How can one account for this finding? The organization of the sensory cortical columns in the cortex consists of excitatory PT neurons and inhibitory interneurons that also mostly interact in a "horizontal" plane in the layered cortex within the same layer of the neocortex. If the forearm part of the S1 receives no sensory input, the neurons in this area can become available for input from neurons in the nearby hand area, which accounts for the fMRI finding of an enlarged hand area. The remarkable finding is that this results in an enhanced perception of the sensory information from the hand with a better two-point discrimination, and this effect is manifested directly after forearm anesthesia. This must mean that the connectivity used for the hand area to expand into forearm area exists normally, but without impact due to the concurrent stronger input from the forearm area.

Similar results have been obtained in the sensory barrel cortex in the rodent. In both the auditory and visual cortices, parts of the tonotopic and retinotopic areas of one part can expand if adjacent areas are not receiving input (Buonomano & Merzenich, 1998). In a sense (so to speak), there is thus a competition between the compartments in the sensory areas for processing units.

5.2.3 Visuomotor Coordination—a Dynamic Process

If you think about it, visuomotor coordination is amazing. It is critical for most types of movement, such as locomotion, reaching, and grasping, or when using tools, like riding a bicycle or driving a car. Most likely, the initial processing of the point in space or object that we like to interact with or the trajectory to follow is processed at an early stage. Subsequently, the parts of the body that will be used for the task in question have to be determined (see also figures 5.5–5.7).

Consider an approaching object, like a ball that you want to catch. This requires rapid calculation of the trajectory of the ball and being able to extend the arms and position the hands in the appropriate spot at just the right moment. The motor command needs to consider not only precise timing, but also the visuomotor details concerning joint angles extending from the eyes to the hand via the arm, neck, and head. This demanding processing takes place in the visuomotor centers in the temporal, parietal, and frontal lobes (Georgopoulos,

1986; Strick et al., 2021; Nakajima et al., 2019). To arrive at accurate and precise motor command, a wealth of information needs to be considered regarding the position of the various parts of the body in relation to each other (and, of course, the surrounding world) and to the object, the ball to catch, and its trajectory in time. The time delay, called "reaction time," is surprisingly small, ranging in the simplest case from 0.1 to 0.2 s to somewhat longer times, when more complex decisions must be made. The reaction time includes processing, decisions, and the coordination of the body with well-timed motor commands.

The motor areas in the frontal lobe in mammals represent the motor-related parts of the neocortex and act via direct projections to the midbrain, brainstem, and spinal cord and, equally important, indirectly via the striatum (Strick et al., 2021; Lemon, 2008; Kaas et al., 2022). For the cortically initiated movements to become behaviorally relevant, they must be adapted to the surrounding environment perceived through the senses, especially vision (see figure 5.7). If one reaches for an object, clearly vision is indispensable, but when moving in a dark environment, hearing is helpful and haptic information can be important, particularly in rodents, with the whiskers being critical when moving along dark underground passages. In mammals, visual information from the retina via the thalamus reaches the primary visual area (V1), which is important for object recognition.

In vertebrates, visual information reaches the cortex/pallium via two avenues: (1) from the retina via the tectum/superior colliculus (SC) to the thalamus (pulvinar) and then to visuomotor areas in the temporal and parietal lobes (Isa et al., 2021); and (2) from the retina via the thalamus (lateral geniculate nucleus, or LGN) to the visual areas in the cortex. Although both pathways exist, the former, which provides preprocessed information in the tectum, appears to be the major pathway for visuomotor coordination in most vertebrates, including mammals, except for primates, where the pathways via V1 takes on a more prominent role (see Kaas et al., 2022, and chapter 3 of this book). Lesions of the tectum in most mammals give rise to severe deficits in visuomotor coordination, whereas lesions to the primary visual areas (V1) have very limited effects (Kaas et al., 2022). In contrast, the converse is true in primates, and lesions of V1 make a primate blind. More prominent projections have evolved from V1 to the higher-order visuomotor areas in the temporal and parietal lobes and further to the frontal lobe–what has been referred to as the "movement-related dorsal stream."

It is remarkable that part of this visuomotor machinery nevertheless is in operation in blind individuals. After a lesion of V1 leading to blindness, a person can nevertheless catch a ball thrown at them, a phenomenon called "blindsight" (Weiskrantz et al., 1974). These individuals then use visual information regarding salient objects moving near them, and they can remarkably either avoid or grasp the object with appropriate visuomotor coordination. Note that this happens even though the individual is not consciously aware of the object that is approaching. This ability depends primarily on processing of visual information gated through the SC and thalamus directly to the higher-order visual areas in the dorsal stream, and then further to the motor centers in the frontal areas (Isa et al., 2021). Individuals with vision, whether humans or other mammals, most likely also use this processing route for visuomotor coordination, although primates also have a prominent contribution channeled via V1.

5.2.4   The Dorsal Stream for Visuomotor Control in Primates

In primates, visual information from V1 is processed along two main pathways (Ungerleider & Mishkin, 1982)—the ventral stream to the temporal lobe for object recognition, sometimes called the "what stream," and a dorsal stream via the temporal and parietal lobes to the motor areas in the frontal lobe, called the "where pathway." To guide our movements and enable us to interact with objects in the environment, an elaborate process needs to take place within the dorsal stream. This information is then forwarded to the premotor areas in the frontal lobe.

Visual input from the V1 and via the pulvinar nucleus in the thalamus (from SC) provides spatial and motion information regarding the location of objects around us and our own movements to an area within the temporal lobe named V6 (primates), in which visual motion in the entire visual field is recorded (figure 5.3). From V6, two main "where pathways" originate, the dorsomedial and the dorsolateral streams (Galletti and Fattori, 2018). The latter pathway analyzes the location of objects in focus for our interest in relation to our own movements. Several compartments, including MT and MST (see

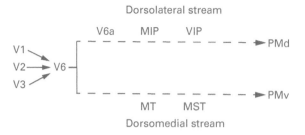

Figure 5.3

The dorsolateral and dorsomedial visual streams forwarding information from the V1–V3 visual areas that feed information to V6, which provides input via the dorsolateral stream to the dorsal premotor area in the frontal lobe and via the dorsomedial stream to PMv. Abbreviations: MIP, middle intraparietal area; MST, medial superior temporal area; MT, middle temporal area; PMd, dorsal premotor area; PMv, ventral premotor area; VIP, ventral intraparietal area.

figure 5.3), within this dorsolateral stream process different aspects of motion in the visual field. For instance, V6/MT analyzes the central foveal aspect (only present in primates; Kaas et al., 2022). The dorsomedial stream (V6a, MIP, VIP) is focused on the spatial location of objects to allow a skilled reaching and grasping of them. Somatosensory and proprioceptive information is also integrated with visual information at the level of the parietal lobe. The degree of activity in V6a neurons, for instance, depends on the orientation of the hand in relation to an object or the shape of the object.

The dorsomedial pathway projects to the dorsal premotor area just in front of the motor cortex, while the dorsolateral pathway talks primarily to the ventral premotor area. This elaborate processing in the centers within the temporal and parietal lobes is indispensable for the motor centers in the frontal lobe, without which they would be unable to steer the movement of the hands and fingers in a meaningful way. In primates, the precise interaction with objects is most elaborate through the control of their individual fingers. We have mentioned mostly reaching and grasping in this discussion, but in real life, visuomotor coordination often involves all parts of the body (consider, for instance, playing tennis). Most vertebrates have an elaborate visuomotor control that may involve the entire body (consider, for instance, cats, dogs, and birds). One cannot help being impressed by a seagull catching food thrown to it, in midair with great precision, or a dog that catches a bone thrown to it.

5.2.5 Visuomotor Coordination during Walking Depends on Cortical Processing

When you move around and correct your step to go onto the curb of a sidewalk or try to kick a soccer ball, you need to estimate the distance to the object and the time it will take to hit the ball or lift the foot to the sidewalk. Marigold and Drew (2017) analyzed this type of coordination by recording neurons in the posterior parietal lobe of the cat. They had the cat walk on a treadmill while an object appeared on the belt at some distance away, and the cat had to lift one paw to overcome the obstacle. They identified two types of neurons—one that became active at a fixed distance to the object, and gradually increased its firing level until the point at which the correction of the limb trajectory had to be made, and another that became active at a certain time before the correction of the paw movement had to be made. When they increased the speed with which the object was approaching the cat, the distance neurons would always start firing at the same distance to the approaching project, and the timing neurons at the time that was required to make contact. This is clearly critical information needed for a precisely timed action. This information is then forwarded to the premotor areas in the frontal lobe to determine which action should be chosen for the corrective movement (Nakajima et al., 2019). These two types of neurons are thought to extract this information from the optic flow, and it is of course necessary for the cat to identify the obstacle. The corrections are mediated to the spinal cord via the corticospinal tract, and lesions of this pathway render the cat unable to place the limb correctly in a complex terrain (Liddell & Phillips, 1944; Beloozerova & Sirota, 1986).

In conclusion, the control of the versatile hand depends entirely on information from the tactile receptors on the fingertips and hand in order to grasp an object in an optimal way. Visual input is critical for preshaping the hand before grasping and to avoid obstacles that would impede the planned movement trajectory. The visual input depends on information (1) mediated directly via the thalamus to V1 and subsequently to the dorsal stream and premotor cortex (PM), and (2) preprocessed information of salient stimuli in the SC that is mediated via the pulvinar nucleus to the higher-order visual areas in the temporal and parietal lobes, and then to the premotor areas. The latter information underlies blindsight, the ability of blind individuals to react to moving objects.

5.3 THE MOTOR AREAS IN THE FRONTAL LOBE OF PRIMATES AND OTHER VERTEBRATES

The presence of motor areas in the frontal lobes had been established in the end of the nineteenth century in mammals, including monkeys, apes, and humans (Lemon & Sherrington, 1917; Penfield & Rasmussen, 1950; Phillips & Porter, 1977; Lemon, 2008; Strick et al., 2021). The frontal lobe of primates contains several motor areas, but there are fewer of them in nonprimate mammals (see figure 5.4; Strick et al., 2021). The motor cortex (M1) is the largest motor area and is divided into subpopulations of pyramidal tract (PT) neurons, each concerned with the control of the hindlimb, trunk, forelimb, and orofacial muscle groups (see figure 5.1). The representation is the largest in areas involved in precision control, like for the hand and oral cavity with jaws and the tongue, and much less to the trunk, although it represents more than half of the body mass. In primates, the motor representation in M1 is prominent, but there are several additional motor areas, which all project to M1, but also to the brainstem and spinal cord. They include the ventral and dorsal premotor areas just rostral to M1, areas on the medial wall of cortex, the supplementary motor area, and three cingulate motor areas. In addition, there is a special motor area for gaze control, the frontal eye field (FEF), which is located just in front of the premotor areas.

In addition to M1, rodents have a smaller but similar motor representation within the frontal lobe, called M2; the somatosensory representation S1, just caudal of M1; and another representation, referred to as S2 in the parietal lobe. The motor representations on the medial wall of the cortex (anterior cingulate sulcus) are absent in rodents and in primates related to anticipation, attention, and motivation. In rodents, the M2 may correspond to the dorsal premotor area of primates, but indications also suggest that there are similarities to the supplementary motor area (Strick et al., 2021).

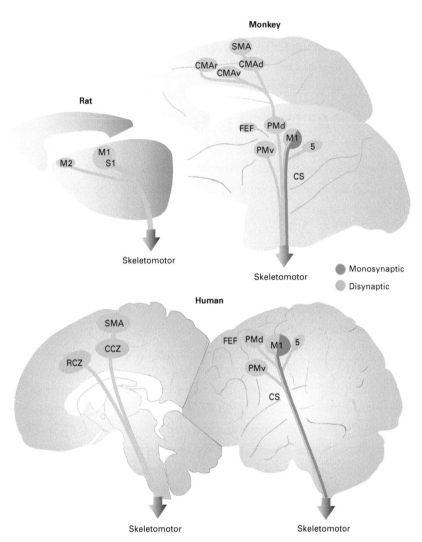

Figure 5.4

Organization of cortical outputs to the spinal cord in rodents, macaque monkeys, and humans. Pathways with monosynaptic access to motoneurons are in red, and pathways with disynaptic access to motoneurons are in purple. The cerebral cortex of the rat has only disynaptic access to motoneurons. This output originates largely from M1, but it also includes smaller contributions from M2 and S1. The cerebral cortex of humans and some monkeys has monosynaptic and disynaptic access to motoneurons. The monosynaptic output originates mainly from new M1, a caudal region of M1 that is largely buried in the central sulcus. New M1 influences motoneurons that innervate proximal muscles, as well as those that innervate distal muscles. Old M1, a more rostral region of M1, is the origin of disynaptic output to motoneurons. In addition, substantial disynaptic outputs

5.3.1 The Motor Cortex (M1)

Projections from the motor areas in the cortex to motoneurons in the spinal cord are mostly disynaptic via interneurons, and they exist in all mammals. In addition, a monosynaptic corticospinal linkage from M1 to the forelimb motoneurons controlling the hand and arm has evolved in humans, apes, and some monkeys such as macaques, but not in animals such as marmosets or squirrel-monkeys. The latter only have the ancient disynaptic pathway, as seen in cats and rodents (Strick et al., 2021). The new monosynaptic pathway in M1 originates from an area just caudal to the old disynaptic M1, which is retained in primates. The monosynaptic corticomotoneuronal pathway (CM) is directed specifically to the motoneurons controlling the hand and its fingers, as well as the proximal arm muscles (Lemon, 2008; Phillips & Porter, 1977).

Neurons in the M1 are activated during reaching movements in different three-dimensional (3D) directions, as shown in the monkey by Apostolos Georgopoulos and colleagues (Georgopoulos et al., 1982, 1988; Schwartz et al., 1988). The monkey was asked to point to a central location and subsequently instructed by a light at any of the eight surrounding points to reach toward this point (figure 5.5). Each M1 neuron activated in the task was broadly tuned to a preferred reaching direction and inhibited by reaching in the opposite direction (Georgopoulos, 1986). However, the neurons are broadly tuned and activated, but to a lesser degree by reaching directions around the preferred direction, which is illustrated in figure 5.5. If the contribution of each of many

Figure 5.4 (*continued*)

originate from multiple premotor areas in the frontal lobe and from a subfield of area 5. One of the major features that distinguish the primate motor system is the proliferation of output pathways from the cerebral cortex, especially from the motor areas on the medial wall of the hemisphere (supplementary motor area, rostral cingulate motor area, dorsal cingulate motor area, and ventral cingulate motor area). The caudal cingulate zone of humans is comparable to the CMAd and CMAv of monkeys. The rostral cingulate zone of humans is comparable to the CMAr of monkeys. The medial wall of the hemisphere is shown in mirror image. Abbreviations: 5, area 5; CCZ, caudal cingulate zone; CMAd, dorsal cingulate motor area; CMAr, rostral cingulate motor area; CMAv, ventral cingulate motor area; CS, central sulcus; FEF, frontal eye field; M1, primary motor cortex; M2, secondary motor cortex; PMd, dorsal premotor area; PMv, ventral premotor area; RCZ, rostral cingulate zone; S1, primary somatosensory cortex; SMA, supplementary motor area. Redrawn and modified from Strick et al. (2021).

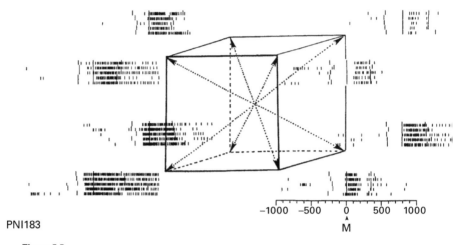

-1000 -500   0   500 1000
              ▲
              M

PNI183

Figure 5.5
The 3D response of a M1 neuron while a monkey is performing a reaching movement from the central position to each of the eight possible positions in each corner of the cube. Note the large response when pointing to the lower left and the inhibition in the upper right, the opposite direction. Courtesy of Schwartz et al. (1988).

recorded neurons were averaged, it was found that as a population, the active neurons formed a population vector, a population code that pointed accurately in the direction of the reaching action that was performed (Georgopoulos et al., 1982, 1988; Schwartz et al., 1988). The degree to which these neurons project primarily to the spinal cord or to brainstem centers like the reaching subset of neurons in the medulla is not known (Ruder et al., 2021).

These center-out reaching tasks are subdivided into two parts: a first ballistic part and a later phase close to the target in which visual feedback can be used to fine-tune the trajectory (Suway and Schwartz, 2019). If the monkeys are asked to generate a more complex shape like an ellipse, they subdivide the task into two components around each maximal curvature, and finally when drawing a figure-eight, they instead use four components, again centered around each peak curvature (Schwartz, 2007). The visual information to the neurons in M1 is mediated via both the dorsolateral and dorsomedial streams in the parietal lobe and then via both the dorsal and ventral premotor areas to M1.

Neurons in the head region of M1 project to the midbrain-brainstem and can take an active part in the control of the motoneurons in the many cranial motor nuclei, and equally important, they affect the many motor centers in the brainstem. Just as important, the M1 activates other descending pathways as the reticulospinal pathways in the pons and medulla. Kably and Drew (1998a,b) investigated the M1 projection to reticulospinal neurons in the cat. The cells from the M1 forelimb area are mainly fast-conducting neurons—one third of which projecting only to the spinal cord, another third giving off collaterals to the pontomedullary reticular formation to activate reticulospinal neurons, and the last third projecting exclusively to reticulospinal neurons. The reticulospinal neurons are activated from the M1 during locomotion, particularly when a cat needs to compensate for an obstacle (a visuomotor compensation). This means that the effects elicited by cortical projections to the spinal cord are complemented by reticulospinal neurons that will further assist in a successful correction of the movement. They also receive complementary input from the cerebellum and other brainstem centers (Matsuyama et al., 2004). Neurons in the PM also project to the reticular nuclei, but their axons are slow-conducting.

5.3.2 The Supplementary Motor Area Can Induce a Sense of Volition

The supplementary motor area (SMA; figure 5.4) is located on the medial side of the hemisphere, just in front of the dorsal premotor area. It is viewed as being upsteam of M1. Much interest has been linked to this structure since SMA neurons are activated prior to movements and earlier than the M1, and the possibility that it plays a role in the initiation of movements (Nachev et al., 2008; Roland et al., 1980a, b). These studies showed that if you ask a person to perform a movement, both the SMA and MI become activated, but if the individual is asked to think about performing the movement, but does not actually perform it, only the SMA was strongly activated; the M1 remained virtually silent. If the SMA and pre-SMA areas are stimulated weakly, the patient reports the urge to move but with no actual movement. A stronger stimulus leads to movement, which the patient reports as if initiated by himself or herself, and thus as volitional. This contrasts with stimulation of the MI, which is perceived as externally imposed (Fried et al., 1991, 2017) and thus not voluntary. SMA can thus induce a sense of ownership of the movement elicited—a sense of volition.

Unilateral lesions of the SMA can lead to difficulties in bimanual coordination. For instance, after this type of lesion, the left and right hands tend to perform the same movement rather than complementing each other as would normally occur. However, if the callosal fibers between the two SMAs were transected, the bimanual deficit disappeared, although a certain clumsiness remained (Brinkman, 1984). In humans, acute lesions of SMA after an operation to reduce the effects of epilepsy can lead to a transient loss of control of the contralateral arm movements, referred to as "motor neglect," and later to reduced bimanual dexterity.

The SMA projects to M1, to spinal cord interneurons, extensively to the striatum, and to the subthalamic nucleus (STN). The SMA receives prominent feedback from the output nuclei of the basal ganglia via the thalamus (Nambu et al., 1996).

To summarize, the SMA is upstream of M1, and its activity is related to the initiation of behavior and to the bilateral coordination of hand movements. It has extensive projections to the basal ganglia and direct projections to the spinal cord as well as to M1. Movements initiated by the stimulation of SMA in humans induce a sense of ownership of the movement—a sense of volition.

5.3.3 The Dorsal and Ventral Premotor Areas, Including Broca's Area

The area just in front of the M1 and lateral to the SMA is referred to as the "premotor area" and is usually subdivided into dorsal and ventral motor areas (PMd and PMv; see figure 5.4). The premotor area can be distinguished both through having different cytoarchitectonic properties and by having neurons with more complex behavior than the M1 neurons. While the latter neurons are active just before and during a given movement, the premotor neurons were active for longer periods before the movement, indicating that they play a role in the planning of movement. Neurons in the PM are heterogenous. They receive prominent input from the parietal lobe, the SMA, and the prefrontal cortex, and they have strong efferent connections to the M1, striatum, reticulospinal nuclei, and spinal cord interneurons (Strick et al., 2021; Schwartz, 2007; Nakajima et al., 2019). Neurons in the PM also project to the reticular nuclei, but their axons conduct more slowly than those from the M1.

Neurons in the PMd are active during reaching and discrete subpopulations are active when the reaching movements have a different orientation (Cisek & Kalaska, 2005). Neurons in the caudal part of the PMv are influenced by a broad repertoire of sensory stimuli such as vision, hearing, and tactile stimuli, mostly related to the peripersonal space. Neurons in the ventral part of PMv are instead concerned with the control of the hand, grasping objects and hand-to-mouth coordination. Some of the neurons in this region have become well known as "mirror neurons," as described by Rizzolatti (di Pellegrino et al., 1992; Rizzolatti & Sinigaglia, 2010). These neurons are thus active if the monkey opens a peanut to eat it, but also if the monkey observes the experimenter opening a peanut with the intention to eat it. The motor act and its intent are thus mirrored in these neurons. Mirror properties are also present in neurons in the parietal lobe that project to PMv.

Broca's area in the left hemisphere of humans partially overlaps with the PMv in nonhuman primates. Lesions of this area lead to different forms of expressive motor aphasia (i.e., the inability to express words and sentences and also to communicate through sign language). Patients understand spoken language and can read, but they have trouble forming words and sentences. Lesions to the corresponding area in the right hemisphere (in right-handers) lead to defects in the prosody, the emotional modulation of sentences that provides a major aspect of communication. Broca's area receives major input from the prefrontal cortex and the temporal lobe, and it is activated during speech in imaging studies, often along with other areas. The precise role of Broca's area in speech production remains elusive and may relate to working memory for syntactic and phonological structure (Friederici, 2002).

From the evolutionary perspective, the presence of mirror neurons in monkeys in the rostral part of PMv that overlaps with the location of Broca's area is of potential interest, as mirror neurons in this area respond to gestures and the meaning of a motor act (e.g., watching somebody manipulating a peanut). This can possibly be extended to the understanding of gestures as used by apes. Broca's area is activated by humans using sign language, so it could be argued that the understanding and execution of gestures in monkeys and apes may be regarded as constituting a precursor of human language.

5.3.4   The FEF: Saccades and Attention to Salient Stimuli

The FEF is located in the frontal lobe (Bruce et al., 1985; figure 5.4) just rostral to the PMd, and it contains relatively large pyramidal neurons in layer 5, which project to the output layer of the superior colliculus (SC) and the gaze centers in the midbrain–brainstem. The FEF receives extensive input from the V1 via the visual centers in the dorsal visual stream (see figure 5.3). Within the FEF, the visual input forms a somewhat distorted retinotopic map. A salient visual stimulus from a given area in the visual field activates a subset of neurons within the retinotopic map in the FEF, and if neurons within this area are activated, a saccade will be generated to direct the eye to this point in the visual field.

The retinotopic map of the FEF is aligned with that of the SC (see chapter 3). Saccades can be generated independently from both the SC and the FEF, although the FEF projects directly to the output layer of the SC. The saccades are formed by the combined action of the downstream vertical and horizontal gaze centers, which are located in the mesencephalon and pons, respectively, which in turn activate the appropriate motor neurons to the eye muscles. After a lesion in either the FEF or the SC, saccades can still be generated, but not if both structures are inactivated (Isa et al., 2021). FEF and SC can thus independently generate the command for the gaze centers to produce an appropriate activation of the vertical and horizontal gaze centers. Normally, however, they most likely work in concert since FEF projects to the output layer of the SC.

The FEF have traditionally been linked to the motor aspect of saccade generation, but now it is also thought of as a center for selective attention. When a salient stimulus occurs somewhere in the visual field, this information is efficiently transmitted through the dorsal stream to the retinotopic map in the FEF, from which a command to elicit a saccade is generated to allow the eye/brain to inspect the salient stimulus (Isa et al., 2021). As discussed in chapter 3, express saccades, which have a shorter latency, can also be elicited at high levels of attention. These are most likely elicited by direct retinal projections to the SC, and then to the brainstem gaze centers.

There is another structure, the basal ganglia, that is critical for the generation of saccadic eye movements (see chapter 4). Under resting conditions, the substantia nigra pars reticulata (SNr) output neurons of the basal ganglia are tonically active and keep the SC output neurons inhibited, which are the same neurons that FEF efferents converge on (Hikosaka & Wurtz, 1983; Kim & Hikosaka, 2015). FEF projects massively to the striatum and is therefore able to exert a fine control over the SNr inhibition of the SC. When the command for a saccadic eye movement is initiated, the direct pathway striatal projection neurons become activated, which inhibits the neurons within SNr that project to the SC (chapter 4). This leads to a removal of inhibition from the output neurons of the SC that mediate the saccade to the downstream motor center, and thereby a facilitation of the saccade-generating circuitry.

In conclusion, in the FEF, there is a retinotopic map from which saccadic eye movements to various points in the surrounding environment can be elicited. There are projections from the FEF to the gaze centers in the midbrain and brainstem, as well as to the output cells of the SC from which saccadic eye movements can be elicited. The FEF are also linked to the mechanisms of selective attention, and when a salient object appears in the visual field, the first thing that normally happens is to elicit a saccade so that the projection of the object is brought into the fovea centralis of the retina, the area with the highest visual acuity.

5.3.5 Long-Lasting Focal Electric Stimulation of the Parietal Cortex, M1, and Premotor Areas Uncovers Separate Functionally Interacting Domains Controlling Various Aspects of Integrated Motor Behavior

The traditional way of studying the effect of microstimulation of the M1 is to use brief stimuli to observe twitches in specific muscle groups or synaptic potentials in the motoneurons. Therefore, it has been possible to address the specificity of different areas in the M1 and at the same time evaluate if the linkage to motoneurons was disynaptic, as in most mammals studied, or partially monosynaptic, as in humans, apes, and some monkeys. Graziano and colleagues (Graziano & Aflalo, 2007; Graziano, 2016) instead used a 0.5s stimulation and could then elicit integrated arm movements such as reaching to grasp, hand to mouth, the orientation of the arms in different positions, chewing, and defensive reactions (figure 5.6).

These findings and the relation to the posterior parietal cortex (PPC) as part of the dorsal stream have been further analyzed in considerable detail by the Kaas's laboratory in a series of studies (Stepniewska et al., 2014, 2018,

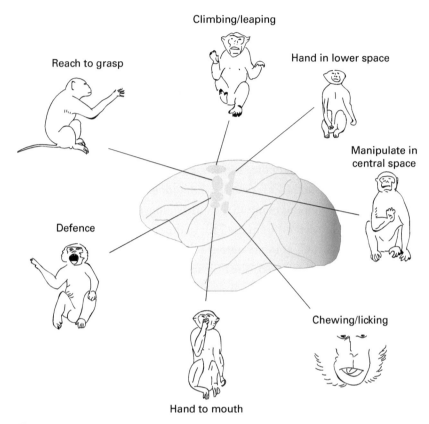

Climbing/leaping

Reach to grasp

Hand in lower space

Manipulate in
central space

Defence

Chewing/licking

Hand to mouth

Figure 5.6

Long-lasting stimuli of the M1 leads to area-specific integrated movements. Intracortical stimulation for 500 ms evoked complex, ethologically meaningful actions. Different actions were evoked
from different cortical zones. Hand-to-mouth actions involved shaping the hand into a grip posture,
orienting the grip toward the head, moving the hand to the mouth regardless of starting position,
opening the mouth, and turning the head to align the mouth to the hand. Defensive actions involved
closing the eyes, pursing the skin around the eyes, folding the ear against the head, turning the
head, shrugging the shoulder, turning the torso, and raising the hand to a blocking posture.
Reaching involved movements of the torso, shoulder, and arm to project the hand, pronation of
the forearm to orient the hand, and postures of the hand as if to preshape for grasping. Climbing
and leaping postures involved bilateral positions of the legs and arms, with the feet and hands
partly curled as if in preparation to grasp branches, and sometimes deflection of the tail as if to
maintain balance during locomotion. Hand-in-lower-space actions involved postures of the arm
that brought the hand to the space near the feet or laterally to the side of the feet and often brought
the hand to a palm-down posture as though to brace the body's weight against the ground.
Manipulation in central space involved complex positions of the wrist and fingers that resembled
typical actions during the manipulation of objects, and often movements of the arm that brought
the hand to the central space in front of the chest or stomach where monkeys typically manipulate
acquired objects. Chewing and licking movements were evoked from the classical primary motor
oral area. Redrawn and modified from Graziano (2016).

2020; Kaas et al., 2022). The caudal part of the PPC has mostly visual input, while the rostral PPC is multisensory and includes somatosensory input. The studies were able to show that the rostral PPC contains seven separate domains, each involved with specific aspects of behavior. From lateral to medial in the PPC, there are separate domains for the face, defensive arm movements, grasping, hand-to-mouth movements, a defensive area, reaching, and most medially a climbing and running domain including forelimbs and hindlimbs (whole-body movements). They showed that neurons within each of these domains projected to separate areas within both the PM and the M1, in which the corresponding behavior could be elicited by microstimulation (see figure 5.7a). Conversely, if the domains within the PM or M1 were inhibited by injection of a γ-aminobutyric acid (GABA) agonist, the behavioral effect of the PPC microstimulation was blocked, suggesting that the PPC effects were channeled through the PM and M1 domains. The caudal PPC with visual input projects to the FEF. This organization can be regarded as further parcellation of the PPC as part of the dorsal stream that has become subdivided into functional compartments or domains, in which each talks with the corresponding domains in the PM and M1.

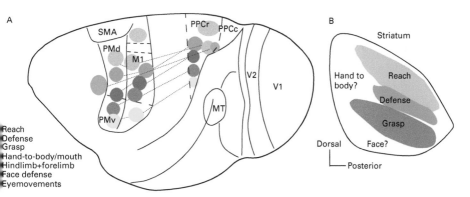

Figure 5.7
Functionally related subdivisions for visuomotor coordination in the posterior parietal, premotor, and motor cortices and striatum. Abbreviations: M1, primary motor area; MT, medial temporal area; PMd, dorsal premotor area; PMv, ventral premotor area; PPCc, caudal posterior parietal cortex; PPCr, rostral posterior parietal cortex; SMA, somatomotor area; V1, primary visual area; secondary V2, visual area. Redrawn and modified from Kaas et al. (2022).

Neurons within the domains of the PPC, PM, and M1 project to the dorsal striatum and the functional subdivisions are maintained (figure 5.7b) so that separate rostrocaudal columns of cells are activated by reaching, defensive reactions, grasping, and other actions. This shows how linked the cortical circuits are to each other, to the basal ganglia, and to downstream motor centers. Whereas the basal ganglia work through disinhibition of downstream motor centers, the cortical domains can provide direct excitation of these motor centers (see also chapter 4).

5.3.6 The Projection Pattern from the Motor Areas of the Frontal Lobe to the Midbrain and Brainstem

In the lamprey motor area in the dorsal pallium (corresponding to the neocortical motor area), there are monosynaptic projections to the output neurons of the tectum/SC, the midbrain tegmentum, the reticulospinal projections, and other brainstem motor areas (Ocana et al., 2015). This is in principle the same projection pattern as in mammals, but it has orders of magnitude fewer projection neurons. It suggests that this basic organization of the motor projections from the cortex has been conserved throughout vertebrate phylogeny.

In mammals, the FEF projects directly to the output neurons of the SC and can thereby induce precise eye movements in specific directions (see chapter 3) via the horizontal and vertical gaze centers. This is important for our ability to rapidly direct the gaze to any salient point in the surroundings—selective attention. In parallel, we have the FEF effects exerted via the basal ganglia, leading to disinhibition (Hikosaka & Wurtz, 1983) of the SC that together control the gaze. These circuits will contribute to the related orienting movements of the head/body and conversely visually evoked evasive movements that are needed to avoid a collision or a threatening stimulus (Kardamakis et al., 2015).

A given PT neuron may give off collaterals to many targets on its passage downstream, which may include the thalamus, midbrain, and brainstem structures. Very powerful monosynaptic projections from the frontal motor areas target the various reticulospinal neurons in the pons and medulla that mediate postural and locomotor corrections, and also target the medullary centers for reaching and grasping. Other parts of the motor areas target the chewing central pattern generator (CPG), and others control the tongue and the licking CPGs (see chapter 2), as well as the jaw muscles. PT neurons also project to the nucleus ruber in the mesencephalon, which contains not only the rubrospinal projections, but also subsets of neurons projecting to the inferior olive and to the medullary centers for reaching (Fidelin & Arber, 2022).

5.3.7 The Specific Role of the Corticospinal Tract: Effects of Lesions

In a very influential study by Lawrence and Kuypers (1968a, b), the researchers made a selective lesion of the corticospinal pathway at the lower brainstem level, leaving the cortical control of the brainstem intact. They were then able to show that the monkeys were able to move around with ease, climbing on the cage walls with no distinct motor deficits, except for the control of the individual fingers. They failed at a task of picking up a peanut from a small hole that required a grip with two fingers, such as between the index finger and the thumb. The monkeys did not recover this ability. The possibility to independently control the fingers may rely preferentially on the CM pathway and allow the versatility of the primate (and in particular the human hand). The monkeys were still able to flex or extend the fingers together to grab or release objects.

Subsequently, in these studies, a selective lesion of the spinal cord was added that removed the descending control from the rubrospinal pathway and some lateral reticulospinal projections. The result was more dramatic—the monkeys lost independent control of the hand in relation to the arm, but they could still locomote and climb. When eating, they had to use the whole arm as a unit and extend it to grab food and bring it to the mouth by flexing the limb. That is, a corticospinal lesion led to an inability to use the fingers independently, while after the rubrospinal and reticulospinal lesions, they were unable to individualize hand control and thus had to rely on a coarse, whole-limb synergy that was very similar to that used in the forelimb during locomotion.

A different pertinent set of experiments was done in the cat. After a lesion of the pyramidal tract (Liddell & Phillips, 1944), cats readily moved around when walking on a flat surface, but they failed miserably when they had to place the limbs accurately, as when walking on a ladder. To place the paws accurately on the rungs of the ladder requires good visuomotor coordination, and an intact corticospinal tract. Recordings from PT neurons show that they are strongly modulated during precision walking in contrast to overground locomotion or when an obstacle impedes the trajectory of the limb during walking (Beloozerova & Sirota, 1986; Drew et al., 2008). The visually related input to the PT neurons in the motor cortex originates from neurons in the parietal lobe. The accurate foot-placement command depends on cortical modulation of the locomotor-related signals from the spinal locomotor generator. Drew et al. (2008) also showed that the same PT neuron was activated both when correcting the foot position in precision walking and in an independent reaching task toward a visually identified object (Yakovenko & Drew, 2015). Under these conditions, the same PT neuron is thus recruited for the same type of movement performed in two very different functional contexts.

Georgopoulos and Grillner (1989) argued that many arboreal species, such as squirrel monkeys and squirrels, throwing themselves from branch to branch during locomotion with great precision and visually evoked reaching movements needed during locomotion, may be the evolutionary origin of the well-controlled independent reaching movements that humans use in everyday life to grasp objects. The capacity to reach out and grasp objects varies widely among rodents, cats, squirrels, and primates, and of course, the need for precision walking is also variable. In mammals in general, the corticospinal disynaptic linkage is most likely responsible for precision walking and reaching. In primates that have evolved the direct CM coupling, the area for the disynaptic control has remained unchanged, presumably serving all types of control except that of independent finger movements.

In conclusion, a selective lesion of the corticospinal tract in primates and cats leads to deficits in fine control of the fingers in monkeys and precision walking in cats, whereas the rest of the motor control remains relatively unaffected. The forebrain control of the midbrain and brainstem is thus sufficient to coordinate most of the motor repertoire, even without direct corticospinal projection. In primates, this includes the projections to the spinal cord from the M1, as well as the projections from the premotor, supplementary, and cingulate areas.

5.4 NEOCORTICAL ORGANIZATION AT THE CELLULAR LEVEL AND THE INTERACTION BETWEEN THE FRONTAL MOTOR AREAS, STRIATUM, AND DOWNSTREAM MOTOR TARGETS

5.4.1 Cellular and Columnar Organization of the Neocortex

The neocortex has six layers and is divided into ontogenetically defined minicolumns with few neurons (Rakic, 2008). These minicolumns do not match the functionally defined cortical column, which may include several minicolumns. Vernon Mountcastle was the first to show that individual cortical columns in the somatosensory cortex have discrete innervation fields, such that the cells in one column are activated only by a specific cutaneous area, or even cutaneous sensory modality, and adjacent columns have a distinctly different sensory input (Mountcastle et al., 1957). The extrinsic input to a column originates from the thalamus and targets layers 3–4 from short- and long-range projections within the cortex. The processing within a column is then signaled to efferent target areas within the cortex, striatum, and other downstream centers. In the visual area, these findings were further developed by David Hubel and Torsten Wiesel (1969), who established the presence of orientation columns (which respond to the orientation of an object as it is moved) and a more elaborate processing of the visual scene with simple, complex, and hypercomplex cells, providing information on the qualities of objects. This work has then been carried further with definition of color and ocular dominance columns.

Initially, it was assumed that the neocortex consists of several stereotyped processing columns with the same cellular composition. It has now become clear that the functional columns can vary in size (e.g., the rodent barrel cortex), neuronal density, and expression of ligands and metabotropic receptors in various neocortical areas. Moreover, the cortex varies markedly between species and the depth of the cortical layers is larger in primates than mice (Jones, 2000; DeFelipe et al., 2002; Herculano-Houzel et al., 2008). Nevertheless, the general organization within the neocortex is at first approximation stereotyped with various types of excitatory pyramidal neurons in the layers, a varied subset of GABAergic interneurons, and intrinsic connectivity within the column.

The fast dynamic interaction within a column at the millisecond level is handled by glutamatergic and GABAergic synaptic transmission, but the responsiveness within a column can also be markedly modified by modulatory input, acting with a longer delay, mediated by metabotropic receptors. There is a varied and rich innervation of different parts of the neocortex by dopaminergic, 5-HT, noradrenergic, and cholinergic neurons that can change the intracolumnar processing markedly. The contribution of each column to the operation of the brain is determined by its projections within and beyond the cortex.

For the motor cortex, the efferent neurons are located in layer 5, which contains the PT neurons, which target downstream motor centers and the striatum; and the intratelencephalic (IT) neurons, which project to the ipsilateral and contralateral striatum and within the cortex. Both IT and PT neurons have extensive apical dendrites that reach the uppermost molecular layer, and they can sample information at all layers of the neocortex. They also have local axonal arbors. PT neurons can excite other PT neurons in the vicinity, while IT neurons can excite both PT and other IT neurons. The input from the thalamus mainly targets layer 4, but also layer 3, and is forwarded to layer 2, and then to layer 5. Layer 6 contains neurons that project to the thalamus, and it emphasizes the dynamic interaction with the thalamus (input/output) and contributes to the operation of the thalamus in terms of how much information is transferred to the cortex and striatum from the thalamus—partially a gating function. The long-range interaction within the cortex is mediated via the molecular layer. It is noteworthy that important processing occurs in layer 1 within the tips of the apical dendrites, far from the soma of the PT and IT neurons, which contributes to complex/cognitive functions such as the perceptual thresholds in the somatosensory cortex (Takahashi et al., 2016). Inhibiting the apical dendrites of the barrel cortex (activated from the whiskers) increases the perceptual threshold to sensory stimulation, while a depolarization instead lowers the threshold. These data clearly show a prominent role for the processing within the apical dendrites.

Most neurons within the cortex are excitatory and glutamatergic (80–90 percent), and the remainder represent many subtypes of GABAergic interneurons specialized for different roles, such as the basket, tufted, Martinotti, and chandelier cells (Markram et al., 2004; Peng et al., 2021). They impinge on parts of the somatodendritic membrane of IT and PT neurons and have various membrane properties. It is, however, beyond the scope of this book to discuss the cortical microcircuits and their respective roles in different sensory, association, and motor cortices in detail. It could be a very interesting subject for a separate treatise, however.

5.4.2 Cellular Organization of the Motor Areas: PT and IT Projections

The motor areas in the frontal lobe target the spinal cord directly via corticospinal pathways (PT neurons), and act via ramifications on circuits in the striatum, thalamus, midbrain, and brainstem, which in turn generate different patterns of behavior (figure 5.8). The direct corticospinal projections have been studied the most, often disregarding the equally important processing circuits in the midbrain and brainstem that are engaged in the control of oculomotor, masticatory, respiratory, and orofacial circuits, and those for vocalization/sound production, including speech. In addition, there are relays to the spinal

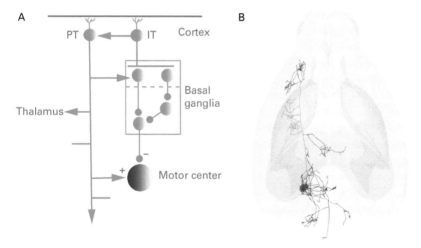

Figure 5.8
Connectivity between cortical PT and IT neurons. *A*. Schematic representation of the connectivity showing that PT neurons can target neurons at many levels, while IT neurons excite PT neurons and have a massive projection to the striatum. Color-coding: excitatory neurons (red) and inhibitory neurons (blue). *B*. Illustrates one example of a reconstructed PT neuron, showing extensive ramifications within the cortex, to the striatum (red) and further to the thalamus, midbrain, and brainstem.

cord via the midbrain/brainstem level, such as the cortico-reticulospinal or cortico-rubrospinal pathways.

The motor areas of the cortex operate in parallel via two different avenues (figure 5.8) as follows:

- Via direct projections from the pyramidal tract neurons (PT), located in the cortical layer 5b, to the striatum, and the downstream motor centers in the midbrain, brainstem, and spinal cord.

- Via the intratelencephalic neurons (IT) in layer 5a of the same motor areas of the cortex as PT neurons, that project massively to the striatum (see chapter 4). Each IT neuron has a broad termination area. In contrast to PT neurons, IT neurons have extensive axonal projections to the contralateral side (see figure 4.10) and ramify extensively bilaterally in the striatum and cortex.

- IT neurons excite PT neurons, but there is no reciprocal connection. IT neurons can thus exert a major effect on the excitability of striatum, and at the same time on PT neurons. IT neurons do not project further downstream. The basal ganglia, via its output level, the SNr and globus pallidus interna (GPi), will then act on downstream motor centers at the midbrain-brainstem level to either disinhibit or inhibit (see figure 5.8).

These two pathways may then act in concert. While the PT neurons excite a given motor center directly, the basal ganglia act through disinhibition of the same motor center, mediated via SNr/GPi. Both pathways can thereby promote the same action (see chapter 4). This general organization applies throughout vertebrate phylogeny, as established in two evolutionary extreme vertebrate groups, the lamprey and mammals (Ocana et al., 2015; Grillner et al., 2020).

5.4.3 Simple and Complex Reaction Time in Humans and Rodents

When a human is asked to press a button as a response to a flash of light or an auditory signal, the minimal latency between signal and the mechanical response is a bit over 0.1 s. This reaction time is the required minimal processing time, which includes the perception of the signal. This provides identification that the subject must recruit the motor system to activate the

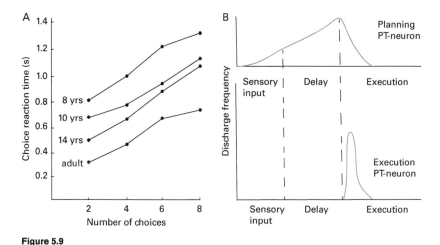

Figure 5.9

Choice reaction time and age in humans and motor planning in mice. *A.* The choice reaction time from two to eight choices is increasing dramatically. It is represented versus the age of the subjects from 8 to 10 to 14 years of age, and adults. It is noteworthy that at the age of 8, the reaction time is approximately three times that of the adult, which has important implications when considering children in urban traffic situations. Modified from Grillner (2012). *B.* During the planning of a movement in the mouse as a response to a stimulus of the vibrissae, the planning type of PT neuron in the tongue motor area in the frontal lobe starts to be active during this phase with sensory activation. The activity is maintained or increases during a delay period (without sensory stimulus) when the mouse must hold the intention to move in working memory. Finally, when the signal to execute comes, the planning neurons are silenced, while a different set of execution neurons become activated. Redrawn based on results from Economo et al. (2018).

appropriate motoneurons and then send the signal to the muscles and make them contract and press the button. If there is a choice between two signals and a requirement to press one or the other button depending on the signal, the reaction time becomes much longer and more complex. The more complex the task, the longer the required time will be, presumably due to a longer decision time. Figure 5.9a illustrates how the choice rection time varies with the number of choices and age. For an eight-year-old, the choice reaction time is more than three times that of adults (!!), and there is a progressive shortening until the teens. Since the length of the reaction time cannot be reduced by training, it means that an eight-year-old would require three times longer to evaluate a possible threatening situation, such as in a traffic situation when a car approaches.

It is critical to consider the neural bases for the reaction time and the decision process, which have been addressed in monkeys and mice with different strategies. In both cases, there has been a choice between one or several signals to respond to, and then the animal must hold the decision for several seconds and then receiving a go signal after which the animal executes the response—right or wrong. One interesting experimental model has been developed by the Svoboda laboratory (Economo et al., 2018; Daie et al., 2021; Peron et al., 2020; Finkelstein et al., 2021). In these studies, they trained mice to move the tongue to the left by activating one set of vibrissae (whiskers), or move the tongue to the right when activating another set of vibrissae. They identified two sets of PT neurons in different parts of layer 5 in the part of the frontal cortex that controls tongue movements. The first set was active during the period in which the animals made the decision and was then remembering the decision until they received the go signal, resulting in the response (figure 5.9b). These neurons increased their discharge gradually until the go signal arrived. It was noted that distractors that could perturb the decision were more efficient in the early period, and the more time that passed, the more robust was the decision during the planning phase. When the go signal came, another set of neurons became active that were responsible for the execution of the tongue movement by activating the brainstem licking CPG. The two types of neurons were in the same general location in the tongue motor area. The first set of PT neurons was engaged in the planning phase and interacted primarily with the thalamus. The other set was engaged in the execution of the tongue movement.

One interesting question was asked subsequently: what happens during the planning phase with the increase in discharge of the planning neurons and the progressively increasing robustness of the decision of the choice that had been made? The sensory input comes via the barrel cortex, the special part of the somatosensory cortex that deals with input from the different vibrissae, with one barrel for each major vibrissa. Did the robustness develop in the barrel cortex or in the tongue motor cortex? Svoboda and colleagues elegantly analyzed this by introducing light-sensitive molecules (channel-rhodopsin (ChR)) into the input layer of the barrel cortex, and by weak stimuli, they could train the mouse to respond to brief light pulses to the barrel cortex with tongue movements in the preferred direction. In this way, they reduced the circuit to the barrel cortex and the tongue motor area, and they showed that the progressive robustness of the decision during the planning phase took place in the motor area. The exact circuitry and cellular mechanisms involved are not yet clear, although a reverberating circuit between the neurons active in the planning phase, some of which project to the thalamus and presumably back, would appear to be plausible candidates. Cellular properties such as plateau potentials could contribute as well and excitatory interaction within nearby neurons in the cortex reinforcing the plateau.

5.4.4 Top-Down Analyses of Motor Actions in Humans: "Voluntary Control"

All movement that are self-initiated must be classified as volitional/voluntary, whether playing the piano, tightrope walking, or just leisurely walking along a street, although the origin of the command may be in different parts of the nervous system (Grillner & Wallén, 2004). As we discussed earlier about the supplementary cortex, an activation of the SMA in humans can lead to an urge to move, and with stronger stimulation, a movement occurs that the patient perceives as a self-initiated, voluntary movement. In contrast, stimulation of the M1 (Fried et al., 1991, 2017) is perceived by the subject as an externally imposed movement. This is thus a marked step in the analyses of voluntary movement that a stimulation of neurons in the SMA can induce a feeling that the human subject actually perceives the movement elicited, as a volitional movement determined by himself or herself rather than the stimulation of the SMA. Similarly, a stimulation of the FEF leads to a saccade that can be perceived as voluntary and related to attention.

When humans speak, several motor-related cortical areas are engaged such as the M1, SMA, PM, anterior cingulate cortex (ACC), superior temporal gyrus, subthalamic neurons, striatum, and Broca's area (Zaccarella et al., 2021). There are many aspects that need to be generated in speech from forming appropriate sentences and recall the appropriate words and pronounce them in a way that is understood by others. Some areas are concerned with syntax and phonemes.

The smallest building blocks are the pronunciations of the vowels. Recordings of single neurons in the human cortex from patients explored for epileptic surgery or subjects of deep-brain stimulation have been very informative (Tankus et al., 2012, 2021). Neurons in the medial part of the frontal lobe were found to be engaged in the control of the shape of the oral cavity, including the position of the tongue. The pronunciation of the vowels "i," "u," "e," and "a" is based on inducing distinctly different positions of the tongue within the mouth—for instance, "i" requires the tip of the tongue to be close to the teeth, whereas "a" requires the tongue to be in a posterior position in the mouth. Tankus et al., (2012) could show that discrete neurons were strongly activated for the pronunciation of each of the vowels. There were thus discrete populations of neurons coding for each vowel, suggesting that there are discrete vowel command units in the brain that can be recruited as we speak. They presumably act via the brainstem microcircuits for tongue control (compare figure 5.9b). These vowel-related neurons may even be activated if one observes the letter in writing, testifying to the link between execution of an action and the perception of the same function.

The link between which circuits in the brain are activated as we perform a movement, as opposed to when we just observe the movement was discussed in relation to mirror neurons in section 5.3.3. Another more complex task in a similar context involves the following experiments (de Manzano et al., 2020). A naive subject was asked to learn to play a simple tune on the piano. As the training was done, one could via magnetic resonance imaging (MRI) observe that an area within the PM was activated. When the tune was presented to the subject later, the same frontal area was activated even though the subject had not attempted to play it. This was specific to the tune that the subject had learned to play on the piano, not to other pieces of music. Again, it seems that the perception of a given tune activates the same area in the cortex as was engaged when playing the melody.

In conclusion, the frontal lobe contains some neurons that give rise to higher-level control and the subjective feeling that an activation of these neurons leads to the perception of the individual that the movement is performed at will. We also have mirror neurons and the processing in Broca's area and beyond that give rise to the spoken word and writing that we are as yet far from understanding.

5.5 MOTOR CAPACITY AFTER LESIONS TO THE NEOCORTEX, INCLUDING THE MOTOR CORTEX

The most radical experiments of inactivation of the neocortex, resulting in a "striatal animal," were performed on a variety of mammals such as rabbits, rodents, and cats in the 1960s and 1970s. These striatal animals could be maintained in the laboratory for periods of months and years and their behavioral capacity investigated. Bjursten et al. (1976) studied cats (see figure 4.2 in chapter 4), in which the entire neocortex was removed a few weeks after birth. As they grew up, the decorticated kittens played with their littermates and were fed by their mothers, and no clear difference in motor proficiency was observed. However, after some months, their interaction with other cats declined and they appeared uninterested in their surroundings. They could react aggressively against other cats and display growling or hissing, the characteristic vocalizations of the cat. They would move around, remember the location of food, and eat and drink normally. Their movement repertoire was not different from that of other cats, and they were able to explore their environment and even find their way out of a labyrinth. Female striatal cats impregnated by intact males displayed normal reproductive motor patterns. As a result, they gave birth to kittens that they nursed appropriately, including feeding over a period of several weeks, and the kittens gained weight as usual. Even a complex pattern of behavior like maternal behavior can thus be performed successfully without the neocortex. This maternal behavior depends primarily on hypothalamic circuits, as later shown in rodents (Kohl et al., 2018; see also chapter 2)

These striatal cats, without any trace of cortex, could thus generate most of the normal behavioral repertoire when it comes to motor coordination. They would walk, assume different characteristic postures, eat, drink, groom themselves, sleep, and express the normal reproductive and maternal behaviors and as kittens, they play well with their littermates. Less information is available about what they were unable to do. As adults, they are described as passive, uninterested, and indifferent to what happens around them, not reacting to external stimuli, such as when somebody is entering the room, which cats normally do. Their "cognitive ability" is clearly reduced. They could discriminate between light and dark and utilize this information for navigation in a labyrinth, presumably based on information from the SC or pretectum in interaction with the striatum. They most likely lacked object vision and the refined sensory processing that takes place in the sensory and higher-order cortices. Although these striatal animals display an advanced behavioral motor repertoire without the neocortex, it seems likely that a detailed analysis of their movements would uncover some specific deficits.

Another set of experiments from the Ölvezcky laboratory with lesions of the frontal motor areas demonstrates the intricate interaction between the motor cortex and striatum (Dhawale et al., 2021). The rats were trained to press a lever twice with a fixed time interval. After training, the rats performed the task well, and if the frontal motor areas were inactivated, they performed it equally well. If the motor cortex was lesioned before the training, however, they could not learn the task. Thus, the motor cortex was critical in the learning period, but not for the performance of the motor task, which depended on circuits in the dorsolateral striatum. The motor cortex presumably played an instructive role during the learning period.

In conclusion, what these experiments show is that with the neocortex absent, but with the basal ganglia and hypothalamus remaining intact, a large part of the standard motor repertoire of an advanced mammal such as the cat can be performed through downstream motor centers. Clearly, movements adapted to the needs of the cat can be initiated and carried out. This impressive subcortical neural machinery can be used to generate the improvised motor repertoire of the intact cat, such as hunting for prey or balancing itself.

5.5.1 Stroke in Humans

Lesions of specialized areas of the cortex often lead to lasting handicaps in humans. For instance, bilateral lesions of the V1 in the visual cortex result in blindness, lesions of Broca's area on the left side (in right-handers, and the reverse in left-handers) lead to an inability to form the sounds required for speech, and lesions of the adjacent arm area result in difficulties in writing. Lesions in the area on the right side corresponding to Broca's area cause an inability to create an appropriate prosody, resulting in a monotonous voice that lacks modulation. This means that the emotional content of the voice is lost, which often can be as important as the spoken words themselves. However, with lesions of this type, the patients are still able to understand spoken and written language.

With vascular lesions in the brain, as with a stroke, it is generally the case that they include not only the cortex, but also subcortical structures such as the striatum, other parts of the basal ganglia, and the many fiber bundles in the capsula interna. The well-known motor symptoms such as paralysis or deficits in the arm and leg movement that occur after stroke are thus in most cases not due to a pure cortical lesion, but the net result of a lesion that includes the underlying subcortical areas. They thus include a deficit in the forebrain control of circuits in the midbrain, brainstem, and spinal cord. In nonhuman primates (as discussed previously), the direct projection to the spinal cord is concerned primarily with delicate finger movements, while most other movements can be executed via circuits in the midbrain–brainstem and other descending pathways.

Fortunately, strokes can now, at least in major medical centers, be treated (if detected very early) by removing blood clots through intravascular catheterization. This impressive development has rescued many patients from a lifelong handicap and saved large costs for society. If, however, a motor deficit occurs after stroke, training should start as early as possible, and it is important to compensate as much as possible for the lost control. The rationale is that the patient should learn to use part of the redundant circuitry available and learn to recruit circuits that previously had not been used to achieve the various motor synergies used to control the arm, hand, and fingers.

Recently developed computer-based rehabilitation schemes in which the patients must actively try to perform an action turns out to be more successful than just training a given set of movements to train, as in traditional physical therapy. A virtual reality–based training protocol using the "Rehabilitation Gaming System" (Ballester et al., 2019; Maier et al., 2019) appears to be efficient. The investigator/doctor can follow the progress of the patient via the internet and will have the ability to modify the task gradually based on the performance of the patient. The gaming aspect may also increase the motivation of the patient, who can monitor the gradual improvement.

5.6 CORTICAL CONTROL OF ROBOTIC ARMS VIA THE BRAIN-MACHINE INTERFACE AFTER SPINAL CORD INJURY

Andrew Schwartz has over several decades analyzed the pattern of neural activity in the monkey M1 and related areas, while the monkey was performing not only reaching tasks (see figure 5.10) but other movements along various trajectories (Schwartz, 1994). The pattern of activity was then interpreted in a population vector algorithm (as discussed previously), in which the population vector could predict the direction of the movement along the path.

With this background, Velliste et al. (2008) took the bold step of recording large numbers of neurons and using this neuronal activity to control a robotic arm via a brain-machine interface. The monkey was first trained to move the robotic arm with a joystick and grasp a piece of food, which served as a reward (Velliste et al., 2008), while the neural activity in the M1 was recorded (more than 100 neurons). In the next step, the arms were prevented from moving (see figure 5.10) and the monkey could thus recruit a neural pattern like the one it had used in the joystick experiments and control the robotic arm that had a shoulder, elbow joint, and gripper as a "hand" that could be rotated to grasp a piece of food. This entailed that the monkey

Figure 5.10

Behavioral paradigm in an embodied control setup. Each monkey had its own arms restrained (inserted up to the elbow into horizontal tubes, as shown at bottom), and a prosthetic arm was positioned next to its shoulder. Spiking activity was processed (boxes at top right) and used to control the 3D arm velocity and the gripper aperture velocity in real time. Food targets were presented at arbitrary positions (top left). Redrawn and modified from Velliste et al. (2008).

visually identified the location of the food, moved the arm in 3D, and, when the gripper was located appropriately, were able to grasp the food and bring it to the mouth. The commands directed to the robot arm were based entirely on the neural activity in the motor part of the monkey's cortex.

With this remarkable result, they went ahead and applied the same thinking to two volunteers with a high spinal cord injury (tetraplegia) who could not move their arms (Downey et al., 2016). Both could learn to control a similar robot arm with the neural activity recorded with a grid of implanted electrodes. In a video I saw, one of the subjects used the robot arm to grasp a bar of chocolate and bring it to her mouth and successfully took a bite out of it. After this, she said smilingly, "One small nibble for a woman—one giant bite for . . ."—most certainly "for science," paraphrasing Neil Armstrong's words when taking his first step on the Moon.

5.7 CONCLUDING REMARKS: THE NEOCORTEX AND THE CONTROL OF MOVEMENT

The telencephalon is critical for our interpretation of the world around us and how we interpret and respond to events—the cognitive aspects in primates, as well as in other vertebrates. The neocortex has particular importance in the context of visuomotor coordination mediated from the V1 via the tempero-parietal dorsal stream to the premotor and motor areas of the frontal lobe and in parallel with the contribution of the SC via the thalamus (pulvinar) to this processing in the dorsal stream. In the posterior parietal lobe, there is multisensory integration subdivided into discrete functional modules related to reaching, grasping, hand-to-mouth, defense, eye movements, and whole-body movements, The output from these modules is forwarded to the corresponding areas in the motor and PM cortices and further to the striatum, still with maintained subdivisions into the same functional modules.

The motor areas of the cortex in the frontal lobe are engaged in the control of the many aspects of movement, including speech and precision movements. They have projections to the numerous motor circuits at the midbrain, brainstem, and spinal cord levels that together represent the behavioral repertoire of a species or an individual (i.e., the motor infrastructure).

The cortical motor areas act through direct projections via the pyramidal neurons (PT) in layer 5 to the motor centers but, equally important, also through massive projections to the striatum via both PT and intracortical projections (IT) that acts through disinhibition of downstream motor centers via the basal ganglia output neurons. Essentially, the functions of the neocortical motor areas and the basal ganglia are intertwined and complement each other. Cortical projections act via direct excitation and through the basal ganglia by disinhibition of the various motor centers.

6 THE CEREBELLUM: CONTRIBUTES TO THE PERFECTION OF COORDINATION

6.1 INTRODUCTION

The role of the cerebellum is different from that of the forebrain, in that lesions of the cerebellum cause a major deterioration of the quality of coordination. However, humans and other vertebrates with major cerebellar lesions can still initiate movement, but the perfection is lost. The coordination breaks down when walking, the precision of reaching and hand movements are lessened, and even speech may become slurred. A traditional test is to ask the patient to move the index finger to the nose. The movement is initiated, but as the finger gets closer to the nose, the finger starts to oscillate, unable to target the nose with accuracy.

Without the cerebellum, movements can still be initiated due to the forebrain control, but without the required precision (Ito, 1984). There are essentially no clear cognitive deficits. The focus of this book is on the forebrain control of initiation and selection of behavior and the midbrain to spinal cord circuits for the execution of movement, but it is important to include an account of the role and function of the cerebellum.

6.2 THE CEREBELLAR CIRCUITRY

The cerebellum, with its characteristic organization, exists in all vertebrates (Eccles et al., 1967; Ito, 1984; Voogd, 1969; Oscarsson, 1976), except the lamprey (Lamanna et al., 2022), which belongs to the oldest group of living vertebrates. The neurons in the cerebellum are densely packed, and it is estimated that the cerebellum has the same number of neurons as the much larger cerebral cortex (Ito, 1984, 1989, 2002; Eccles et al., 1967). The cerebellar cortex is divided into distinct longitudinal zones, each with specific input and output channels to the cerebellar nuclei (Oscarsson, 1976; Voogd, 1969). All the parts of the nervous system communicate with the cerebellum, from the neocortex to the caudal tip of the spinal cord. The input and output are arranged with specific areas for different functions and different parts of the body are represented in an orderly fashion. In a schematic way (see figure 6.1), the vestibular input targets the flocculus, the input from the spinal cord targets the medial and anterior parts of cerebellum (vermis), while the input from the brainstem-midbrain level is located somewhat more laterally at an intermediate level, and that from the cerebral cortex targets the lateral parts, the neocerebellum (figure 6.1). The cerebellum receives detailed information from proprioceptors, exteroceptors such as cutaneous and pain afferents, vision, and hearing. This information is continuously updated and concerned with the motor commands issued (efference copy; Ito, 1984; Orlovsky et al., 1999). The

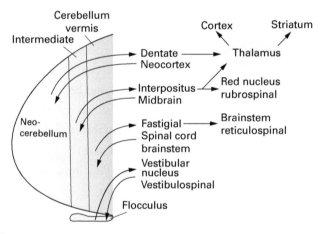

Figure 6.1

Schematic of the afferent and efferent projections of the compartments of the mammalian cerebellum. The compartments of the cerebellum, neocerebellum, intermediate, and vermis have different output channels, the dentate, the interpositus, and the fastigial nuclei, respectively. They in turn project in an ascending direction via the thalamus or in a descending direction via the various descending pathways. For details see figure 6.3.

cerebellum can therefore be regarded as a dynamic storehouse of information of everything that is going on in the motor system at a given point of time.

The only output cells of the cerebellum are the large Purkinje cells, which have a very extensive fanlike (flat) dendritic tree, and they are positioned one after the other in columns of cells (Eccles et al., 1967). Perpendicular to the orientation of the dendritic tree, there are bundles of parallel fibers, the axons of the granule cells, that pass through a sequence of dendritic trees with the potential to form synapses on them (figure 6.2). The fact that the cerebellar Purkinje cells, the output of a major part of the central nervous system, were inhibitory and GABAergic rather than excitatory came as a shock to the neuroscience community, when Masao Ito reported this in the late 1960s (Ito and Yoshida, 1966; Ito, 1972). Now the cerebellum is joined by the basal ganglia, amygdala, and other structures as providing output control through inhibition and disinhibition, while the cortical output, superior colliculus (SC), most descending pathways, and all the sensory afferents have an almost exclusively excitatory output. Ito, at this early stage, also wrote *The Cerebellum as a Neuronal Machine* with Eccles and Szentagothai (Eccles et al., 1967) emphasizing the repetitive neuronal design with parallel circuits.

6.2.1 Climbing and Mossy Fibers Play Distinctly Different Roles

The input to the cerebellum is conveyed via two input channels (figure 6.2):

1. The *granule cells*, which continue as parallel fibers in the cerebellar cortex, are numerous, and thousands of parallel fibers impinge on the dendrites of a single Purkinje cell. Each granule cell receives specific input from only a few mossy fibers (as discussed next) that form very extensive synapses on the granule cells and elicit large excitatory postsynaptic potentials (EPSPs) such that the activation of only a few mossy fibers can lead to an action potential in the granule cell. Each granule cell receives input only from mossy fibers, mostly with very similar function (e.g., a small part of the skin), and thus carry a very precise type of information (Ekerot and Jörntell, 2008; Jörntell and Ekerot, 2011). In addition to Purkinje cells, parallel fibers activate two types of inhibitory interneurons: the Golgi cells, which provides inhibitory feedback to the granule cells; and the basket cells,

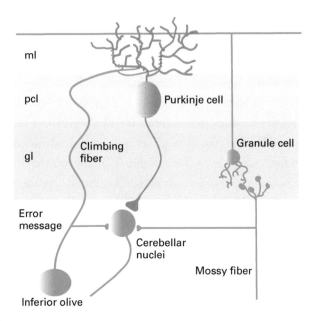

ml

pcl   Purkinje cell

gl   Climbing   Granule cell
          fiber

Error
message

                     Cerebellar
                     nuclei
                          Mossy fiber

Inferior olive

Figure 6.2

Schematic representation of the cerebellar circuitry. The cerebellar cortex has three layers: a granule cell layer (gl), a Purkinje cell layer (pcl), and a molecular layer (ml), in which the parallel fibers from the granule cells extend their axons. The inhibitory (blue) Purkinje cells form synapses on the cerebellar nuclei, which also receive excitation from the collaterals of the mossy fibers that has extensive synaptic terminals on the granule cells. The climbing fibers have very extensive terminals on the entire dendritic tree of the Purkinje cell. The cell bodies of the climbing fibers are located in the inferior olive.

which provide lateral inhibition to Purkinje cells located laterally to those activated from the local parallel fibers (Eccles et al., 1967; Ito, 1984, 1989).

2. The *climbing fibers originating from the inferior olive*. In contrast to granule cells, only one climbing fiber innervates a single Purkinje cell, and its axonal arbor is distributed over the entire dendritic surface of the cell (Ito, 1984). The climbing fiber activation depolarizes the dendritic arbor of the Purkinje cell and leads to a prominent depolarization that can induce a plateau potential in parts of the dendritic tree, which will promote a long-lasting depolarization and synaptic plasticity (Ekerot and Oscarsson, 1981; Ekerot and Kano, 1985). The climbing fibers are mediating an error message to the Purkinje cells and activate the entire dendritic tree and can give rise to

synaptic plasticity in the parallel fiber synapses that provide input to the Purkinje cells at the same time as the climbing fiber signal arrives.

The error message can be from a wound, for instance, as if one hurts a foot, causing a pain response that activates the climbing fibers (Ekerot et al., 1987). If occurring during a movement, it will induce changes of the movement to minimize pain (e.g., limping). For eye movements, a retinal slip over the retina will produce blurring of vision and a climbing fiber response (Ito, 1972, 1984).

The inferior olive is divided into many discrete modules, each with specific input from a given part of the spinal cord, brainstem, or cortex (Oscarsson, 1968, 1976). Within each module, the local inferior olive neurons are connected through gap junctions (Llinás, 2013) and thus tend to become synchronized.

6.2.2 The Organization of the Cerebellar Cortex and the Output Targets

The cerebellar cortex is subdivided into longitudinal zones with specific inputs (Oscarsson, 1968, 1976; Voogd, 1969; Apps et al., 2018) that further subdivide the vermal, intermediate, and lateral parts of the cerebellum into functional stripes. In the vermal and intermediate parts (figure 6.1), furthermore, there are small microregions that have specific inputs from groups of climbing fibers and sets of parallel fibers that impinge on the Purkinje cells. Together, they form what is called a "micromodule" (De Zeeuw, 2021). Each micromodule is thought to control a discrete muscle group, somewhere in the body, via downstream pathways to the motor neurons. There are numerous micromodules, side by side. Micromodules appear to be of two kinds—those in which Purkinje cells have a high resting rate and then reduce their activity during learning, and another group of cells with a low rest rate that instead increase their activity during learning (De Zeeuw, 2021).

The output from the lateral part of the cerebellar cortex, the neocerebellum, is channeled via the dentate nucleus to the thalamus, and then to the basal ganglia and cortex (figure 6.1). There are major connections to the motor areas of the frontal lobe, as well as to other parts of the cortex. As noted earlier, the entire neocortex projects via the lateral reticular nucleus as mossy fibers to the new lateral parts of the cerebellar cortex, and there is thus a reciprocal interaction.

The intermediate part of the cerebellar cortex projects to the interpositus nucleus, which in turn projects to the thalamus and the red nucleus in the mesencephalon, while the medial part of cerebellum, the vermis, targets the fastigial nucleus. Findings by Fujita et al. (2020) have revealed that there is a very detailed compartmental and functional organization with input from specific parts of the cerebellar cortex to five division of the fastigial nucleus (figure 6.3). The vermal area (yellow) projects to a posturomotor section of the fastigial nucleus, which is concerned with the control of reaching/grasping, locomotion, posture, and cardiorespiratory circuits. There is an oromotor area (pink) in a more lateral compartment of the cerebellar cortex, which controls the yaws, and that represents a major control area for exploration, feeding, and defense. It also includes whisking and arousal. These two areas contain large neurons considered to mediate rapid responses and online control. A third area (blue), juxtaposed with the oromotor area, is concerned with orienting responses and targets the SC and is involved in eye/head movement. This area also projects to the thalamostriatal neurons and to the periaqueductal gray (PAG; threats). Surrounding the posturomotor area, there is a positional-autonomic area (green) involved in the coupling between movement and autonomic functions (respiration, circulation). Finally, an area (mauve) with somewhat different characteristics, referred to as "vigilance," is related to defensive arousal, freezing, and the control of dopamine neurons in the substantia nigra pars compacta (SNc), which is associated with salience/reward and motivation. The fastigial neurons in the last three areas are somewhat smaller than in the first two, suggesting a more modulatory role. All these compartments contain glutamatergic projection neurons, but there are also GABAergic neurons within the fastigial nucleus.

The inferior olive projections are also divided into five compartments, which project to the corresponding areas in the fastigial nucleus and the cerebellar cortex. They receive input from the fastigial nucleus, the spino-olivary pathways, and many other structures related to the operation of the five fastigial compartments.

Not only do the dentate and interpositus nuclei project to certain thalamic nuclei, but also many of the functional compartments in the fastigial nucleus target the thalamic nuclei related to motor function. The thalamic

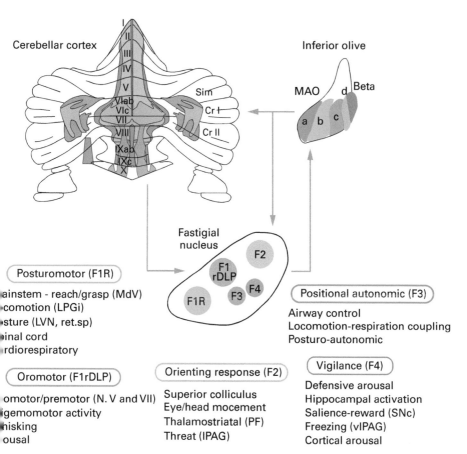

Cerebellar cortex

I
II
III
IV
V
VIab
VIc
VII
VIII
IXab
IXc
X

Sim
Cr I
Cr II

Inferior olive

MAO   d Beta
a b c

Fastigial nucleus

F2
F1 rDLP
F4
F1R
F3

Posturomotor (F1R)
ainstem - reach/grasp (MdV)
comotion (LPGi)
sture (LVN, ret.sp)
inal cord
rdiorespiratory

Positional autonomic (F3)
Airway control
Locomotion-respiration coupling
Posturo-autonomic

Oromotor (F1rDLP)
omotor/premotor (N. V and VII)
igemomotor activity
hisking
ousal

Orienting response (F2)
Superior colliculus
Eye/head mocement
Thalamostriatal (PF)
Threat (IPAG)

Vigilance (F4)
Defensive arousal
Hippocampal activation
Salience-reward (SNc)
Freezing (vIPAG)
Cortical arousal

Figure 6.3

Modular circuit connections of excitatory fastigial projection neurons provide circuit substrates for coordinating five broad organismal functions. Schematics summarize cerebellar modular circuit connections that link distinct types of fastigial nucleus neurons with specific neurons in the inferior olive, cerebellar cortex, and downstream brain regions. Projection targets of each fastigial nucleus cell type are indicated in different colors (as described in the text). Specific functions associated with each collection of projection targets are indicated; proposed broad organismal functions of each module are encircled above. To show the distribution of Purkinje cells associated with each module, a flat map of the mouse cerebellar cortex with vermis lobules is indicated numerically. Inferior olive subnuclei of the caudal medial accessory olive (MAO) are denoted as a, b, c, d, and beta. Circles around each fastigial nucleus cell type indicate its relative size and parasagittal position. Abbreviations: Cr I, Crus I; Cr II, Crus II; IPAG, lateral periaqueductal gray; LPGi, lateral paragigantocellular nucleus; LVN, lateral vestibular nucleus; MdV, medullary reticular nucleus, ventral; N. V, trigeminal nucleus; N. VII, facial nucleus; PF, parafascicular thalamic nucleus; ret. sp., reticulospinal nuclei; Sim, simplex lobule; SNc, substantia nigra pars compacta; vIPAG, ventrolateral periaqueductal gray. Modified with permission from Fujita et al. (2020).

nuclei forward information to the cerebral cortex, including the frontal lobe and to the striatum. The fact that there is a major efficient transmission from the cerebellar nuclei to the intralaminar nuclei in the thalamus, and further to the striatum, has been appreciated only during the last several years (Bostan & Strick, 2018; Caligiore et al., 2017; Chen et al., 2014; but see, however, Strick et al., 2009). This is a very efficient disynaptic transmission from the cerebellar nuclei to the striatum. This connection most likely provides important information to the striatum since the thalamic input to the striatum represents no less than 40 percent of its excitatory input, with the remainder coming mainly from the various parts of the cortex (see chapter 4). What types of movement-specific information that is provided to the striatum from the cerebellum is not yet clear, but since the cerebellum is considered important for the timing of events (Llinás, 2013; Tsutsumi et al., 2020), this could well be an important aspect of the interaction between these two structures engaged in the control of movement.

In conclusion, there is an intricate and precise organization of the fastigial nucleus related to function (i.e., posturomotor, oromotor, orienting movements, positional autonomic, and vigilance), with specific projections to structures in the spinal cord, brainstem, midbrain, thalamus, and inferior olive (Fujita et al., 2020). Each of these fastigial compartments receives input from its own territory in the cerebellar cortex, to which the various parts of the inferior olive project (figure 6.3). These findings emphasize the fine granularity of the processing that takes place within the modules of the cerebellum at the input level, the cerebellar cortex and nuclei, and the projections of the climbing fiber from the inferior olive to the Purkinje cells and cerebellar nuclei.

6.3 SPINAL CORD INTERACTION WITH THE CEREBELLUM: LOCOMOTION AND OTHER MOVEMENTS

The locomotor movements engaging all parts of the body are among the most complex since they need to be adapted to a wide variety of conditions, speeds, terrains, and objects of locomotor activity (e.g., foraging or escape). It is therefore critical that all the circuits involved be optimally calibrated, which is a task for the cerebellum (Orlovsky, 1972a,b,c; Orlovsky et al.,

1999). There are four major pathways that forward different types of information to the cerebellum from the spinal cord concerned with locomotion, but also other rhythmic motor activities (figure 6.4):

1. The *dorsal spinocerebellar tract (DSCT)* originates from the lumbar spinal cord and forwards information from proprioceptors such as muscle spindles and Golgi tendon organs, which become activated during each step when walking, as well as during other movements (figure 6.4). The DSCT neurons therefore become rhythmically active during locomotion (Arshavsky et al., 1972) and forward important information regarding the actual phase of the movement of the limb within the step cycle (hip joint angle) and the length of the limb (dynamic distance between the foot and the hip joint). The DSCT keeps the cerebellum continuously informed of the exact configuration of the hindlimb and whether it is flexed or extended, which is of particular importance when the limb makes contact with the ground at the end of the swing phase (Orlovsky et al., 1999; Grillner & El Manira, 2020). This provides dynamic information regarding the actual ongoing movements, which is transmitted to the cerebellum via the DSCT, which terminates as mossy fiber terminals on the granule cells in the cerebellar posturomotor area (figure 6.3). They continue as parallel fibers in the cerebellar cortex to synapse onto the dendrites of the Purkinje cells (see figure 6.2). The mossy fibers elicit large unitary EPSPs on the granule cells (Ekerot and Jörntell, 2008), and they may each receive only a few afferent DSCT fibers and can therefore forward very specific proprioceptive information to the Purkinje cells.

2. The *ventral spinocerebellar pathway (VSCT)* forwards information regarding the commands issued by the central pattern generator network (CPG) for locomotion in the spinal cord (Arshavsky et al., 1978a, 1984). VSCT neurons are activated parallel to the commands to the flexor muscles. The VSCT also terminates as mossy fibers on a separate set of granule cells from that of the DSCT. The VSCT provides a copy of the commands issued to flexor motor neurons in each step cycle. This is referred to as an "efference copy."

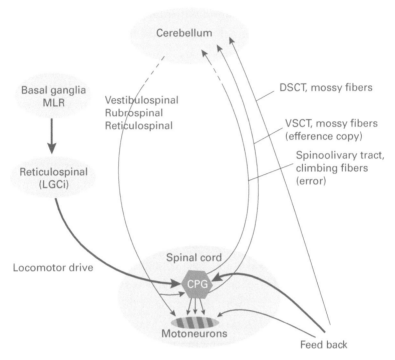

Figure 6.4
Interaction between the cerebellum and the locomotor CPG. The cerebellum receives input from the moving limb through the DSCT, efferent CPG commands via the VSCT, and error information through the spino-olivary tract. The cerebellum provides compensatory adaptations through the vestibulospinal, rubrospinal, and reticulospinal pathways. Abbreviations: LGCi, lateral paragigan-tocellular nucleus; MLR, mesencephalic locomotor region. From Grillner and El Manira (2020).

3.   The *spino-reticulo-cerebellar pathway (SRCP)* provides efference copy information regarding the CPG commands to the extensor muscles (Arshavsky et al., 1978b, 1984). The spinoreticular information is relayed in the lateral reticular nucleus and continued as mossy fibers, to terminate on the granule cells, and then via parallel fibers to the Purkinje cells. Also, propriospinal neurons in the cervical spinal cord, involved in reaching movements have collateral that provides efference copy information to the lateral reticular nucleus and cerebellum via mossy fibers (Azim et al., 2014). The combined information forwarded to the anterior lobe of the cerebellum (posturomotor area) from the spinal cord circuits is concerned with both the commands issued from the extensor and the flexor part of the spinal CPGs through the

VSCT and SRCP, and equally important, with the actual movements that are produced and conveyed through the DSCT. This information is thus able to evaluate if the CPG commands issued have resulted in the anticipated movement or if some perturbation may have occurred (figure 6.4).

4. The *spino-olivo-cerebellar pathway* forwards information of the error message type to the inferior olive, which consists of olivocerebellar neurons, each of which terminates as a climbing fiber on a few selected Purkinje cells (figure 6.2). At rest, the olivary neurons have a very low rest rate (e.g., 1 Hz), but if an intense activation occurs (e.g., from local pain afferents), they become strongly activated and provide an intense discharge over an extended period (Yanagihara & Udo, 1994). The climbing fibers, when activated, produce a complex spike due to massive activation of the dendritic tree of the Purkinje cell, while parallel fiber activation leads only to a conventional, simple spike. If a climbing fiber burst occurs, this leads to a massive Ca^{2+} entry into the Purkinje cell dendrites, and a synaptic plasticity is induced in the parallel fiber synapses that are activated. The underlying activation of sets of protein kinases and phosphatases has been described in many reviews (Ito, 2002) and will not be dealt with here.

6.3.1 Processing in the Cerebellum and Downstream Control of Spinal Circuits

The input from the DSCT, VSCT, and SRCP is fed into the anterior lobe of the cerebellum, and during locomotion, each of these pathways is rhythmically active and activates its own set of granule cells/parallel fibers (Arshavsky et al., 1972, 1978a, 1984; Orlovsky et al., 1999). The DSCT is signaling how the step proceeds within each step cycle based on proprioceptive information, while the VSCT and SRCP become activated along with the flexor and extensor commands from the CPG, respectively. This is mirrored in the parallel fiber activity and the subsequent excitation provided to the Purkinje cells that will become rhythmically active. The GABAergic Purkinje cells represent the only output of the cerebellum.

In the medial part of the cerebellum, the posturomotor area (spinocerebellum), Purkinje cells target the fastigial nucleus, which activates reticulospinal neurons, and Deiters's nucleus, the origin of the fast vestibulospinal tract, providing monosynaptic excitation to extensor motoneurons (see also figures 6.3 and 6.4). In the intermediate part of the cerebellum, the Purkinje cells target the interpositus nucleus, which projects to the red nucleus, the origin of the rubrospinal pathway, and to the thalamus. The net result is that during locomotion, vestibulospinal, rubrospinal, and reticulospinal neurons become rhythmically active. The last two operate with the flexors, while the vestibulospinal pathway is active with the extensors (Orlovsky, 1972a,b,c; Orlovsky et al., 1999). This means that the cerebellar output reinforces the activity of the spinal CPG during ongoing locomotion.

If the cerebellum is inactivated, the modulation of all three descending pathways disappears, which means that that the modulation of the activity of the various pathways depends entirely on the input from the cerebellum (Orlovsky et al 1972a,b,c; Orlovsky et al., 1999). The rubrospinal and reticulospinal inputs remain active without modulation, but with a lower overall frequency of discharge. In contrast, the vestibulospinal input is active at a higher resting rate when the cerebellum is inactivated than when it is intact, which relates to the fact that the vestibulospinal neurons have direct input from the inhibitory GABAergic Purkinje cells.

A perturbation of the locomotor activity (e.g., by delaying the transition from the support to the swing phase) will cause a change in cerebellar activity and the activity of the descending pathways. This means that efference copy information from the CPG, mediated via the VSCT and SRCP, is critical. Interference with the motion of the lower part of the leg has no or little impact on the cerebellar activity. The modulation of the locomotor activity through the descending pathways does contribute to the overall motor activity since lesions or cooling of the anterior cerebellar cortex leads to altered movement amplitude (Yanagihara & Udo, 1994; Orlovsky et al., 1999; Grillner & El Manira, 2020).

6.3.2 Cerebellar Learning and Motor Coordination during Locomotion

The cerebellum can adapt movements over longer periods through a form of learning. This is an error-driven process mediated through the climbing fibers, through which movements (whether eye or locomotor) can be modified. A wound in the foot leads to a limping motor pattern to minimize the pain (Ekerot et al., 1987), which most likely is a cerebellar adaptation. In another example, when a human or a mammal is asked to walk on two belts going at different speeds, one for the left leg and another for the right leg, a modified motor pattern with a complex adaptation regarding both temporal and spatial adaptations occurs (Darmohray et al., 2019). Recordings of the climbing fibers in the anterior cerebellar lobe show that an enhanced climbing fiber activity will occur, particularly in the touchdown phase of the limb movement, representing the most critical phase of the limb trajectory in each step cycle. Lesions of the anterior lobe led to an inability to learn the new adapted motor pattern. Any sudden perturbations occurring during locomotion will lead to a marked increase in climbing fiber activity, providing an error signal.

6.4   THE CEREBELLUM AND THE VESTIBULO-OCULAR AND OPTOKINETIC REFLEXES: CALIBRATION OF MOTOR ACTION

6.4.1   The Vestibulo-Ocular Reflex

Our ability to rapidly focus on salient stimuli through rapid eye movements (saccades), bringing the object of interest into the fovea centralis, the area of retina with the highest density of photoreceptors, is essential for everyday life. The retinal afferents project to a retinotopic map in the SC (see chapter 3) and via the thalamus to the visual areas in the cortex. The eye movements are due to the activation of neurons in the SC and the frontal eye field (FEF) and are related to attention (see chapter 5). To maintain the gaze at the same point while the head is moving requires rapid compensation. For example, if you are running with your head moving up and down in each step cycle or when being transported in a horse and carriage, the head will move up and down in an unpredictable way, which would cause the vision to become blurred, if not for compensatory eye movements. The movement of the head in space, however, is constantly monitored by a vestibular apparatus (i.e., a semicircular canal), which provides well-calibrated corrective signals in all planes to activate the appropriate combination of eye muscles to counterrotate the eye so that the focus on the object is maintained by keeping the image within the fovea centralis. For our survival, this aspect of vision, which allows for rapidly inspecting salient stimuli, friend or foe, is critical. Maintaining the gaze is handled by the vestibulo-ocular reflex that is present in all vertebrates (Wibble et al., 2022).

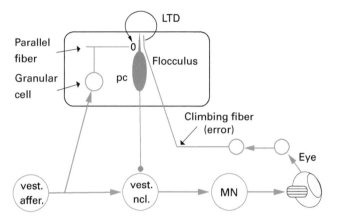

Figure 6.5

Schematic representation of the vestibulo-ocular reflex and its adaptation through the cerebellum (flocculus). The vestibular afferents activate granular cells that continue as parallel fibers and make synapses onto the Purkinje cells (pc). Retinal afferents activated by a retinal slip provides an error message forwarded by the inferior olive as a climbing fiber message that elicits a massive depolarization of the Purkinje cell dendritic tree. This leads to an accompanying Ca^{2+} entry with effects on downstream protein kinases leading to a modification of the parallel fiber synapses onto the Purkinje cell (LTD), but only the synapses that are activated concurrently with the climbing fiber input. Abbreviations: MN, motor neurons; vest. ncl., vestibular nuclei; vest. affer., vestibular afferents.

The basic organization of the vestibulo-ocular reflex is provided by vestibular afferents connected to vestibular interneurons that impinge on the motoneurons (figure 6.5) that activate the eye muscles, which have the fastest muscle fibers in the body, to provide rapid correction of eye position. Take, for instance, a rapid horizontal movement to the right, which would require a counterrotation to the left of both eyes but with opposing eye muscles (lateral rectus on one side, and medial on the other).

It follows that the vestibulo-ocular reflex needs to be precise, and not overcompensate or undercompensate, a calibration task for which the cerebellum is required. If you wear new glasses or are subject to even larger deviations by seeing the world through prisms, the vision gets blurred. This will give rise to an increased activity in the climbing fibers, resulting in an error signal that acts on the level of the Purkinje cell dendrites and readapts the strength of the parallel fiber synapses onto the Purkinje cells located in the vestibular part of the cerebellum, the flocculus. That the climbing fibers act as error detectors and can modify the gain of the parallel fiber synapses was first conceived by Ito (1972) and is still one of the main principles of cerebellar function. The climbing fiber activation leads to calcium ions entering the dendrites of Purkinje cells, which in turn activates a complex set of protein kinases that can elicit long-term depression (LTD) of the parallel fiber synapses that become activated by the vestibular input in conjunction with the climbing fiber activity (Ito, 2002). This will result in a modified degree of activation of the GABAergic Purkinje cells, and therefore a modified gating of the transmission in the vestibular nuclei. However, additional plasticity (LTP or LTD) occurs in other synapses, notably in those of the mossy fibers that provide collateral activation of the floccular target neurons through cerebellar Golgi interneurons, on the cells of the different output nuclei, and in the cerebellar cortex (Lisberger, 2021; De Zeeuw, 2021).

It is easy to understand how important the vestibulo-ocular reflex is in everyday life and for survival, and therefore why an accurate cerebellar mechanism for recalibration is crucial.

6.4.2 The Optokinetic Reflex

The optokinetic reflex also provides another mechanism for stabilizing the vision. If the head remains in a stable position, while the object being looked at moves over the retina rather than remains in a stable position, it is referred to as a "retinal slip" (Lisberger, 2021). This gives rise to a retinal afferent response via the pretectum and the vestibular nuclei and the gaze centers that activates the eye muscles. This is a conserved subcortical circuit in which the optokinetic reflex converges partially on the same neuronal optokinetic response over the retina (Wibble et al., 2022). It has a much longer latency than the vestibulo-ocular reflex with its very short latency. One can easily convince oneself of the marked difference by holding a finger in front of oneself while rapidly rotating the head. Through the vestibulo-ocular reflex, one can maintain a stable image of the finger even at high head velocities. In contrast, if you instead move the finger back and forth, keeping the head stationary, one loses the image at a very modest speed of rotation! Whereas the optokinetic reflex is subcortical (Wibble et al., 2022), a reoccurring retinal slip leads to a climbing fiber response and the same type of cerebellar recalibration takes place as outlined for the vestibulo-ocular reflex (Ito, 1984). This will help to recalibrate the optokinetic reflex, and thereby eye movement.

6.5 PARALLEL FIBER SYNAPSES ONTO PURKINJE CELLS: ACTIVE AND SILENT SYNAPSES—PLASTICITY

The parallel fiber synapse onto Purkinje cells and cerebellar interneurons can both be downregulated as in LTD or upregulated as in long-term potentiation (LTP). LTP can occur when parallel fibers are stimulated in an LTP protocol, while climbing fiber activation tends to cause LTD (see, De Zeeuw, 2021). In the anterior lobe of the cerebellum, the skin surface is represented in certain sagittal zones so that each Purkinje cells may have very limited input from one digit or a very localized skin area limited to a few cubic centimeters. Light cutaneous activation is sufficient to activate the receptive field (Jörntell and Ekerot, 2011). The activation is then mediated by a discrete set of mossy fibers/parallel fibers, leading to an activation of Purkinje cells.

Of all the synapses that impinge on the dendritic tree of a Purkinje cell, most remain silent, but they can be awakened within minutes. A brief electrical repeated stimulus of the surrounding skin areas, with trains of 5 minutes, elicits a marked expansion of the receptive field within minutes, particularly when combined with climbing fiber input. The receptive field can then be markedly expanded from a very small area to include the whole limb. This is due to synaptic plasticity in the synapses between parallel fibers and the Purkinje cells and interneurons (Jörntell and Ekerot, 2011). This example shows that the potential receptive field is flexible and subject to modification due to the input being received from climbing and parallel fibers (De Zeeuw, 2021). The implication is also that, of the many parallel fibers that impinge on a Purkinje cell, only a few have active synapses from a given area; but with specific additional climbing fiber input, other synapses can be recruited, resulting in a modified receptive field. A given Purkinje cell thus has the potential to serve different innervation fields that depend on changes in the gains of the various parallel fiber synapses. It is estimated that perhaps only 2.5 percent represent active parallel fiber synapses and the rest remain silent, but with the potential to be recruited rapidly into action (Jörntell and Ekerot, 2011). Each Purkinje cell thus has the potential to add or subtract new active synapses.

6.6 THE CEREBELLUM'S ROLE FOR LEARNING TO ASSOCIATE TWO RELATED BUT INDEPENDENT PROCESSES: CONDITIONED REFLEXES

One aspect of learning involves the calibration and optimization of the performance of various types of movements, which is clearly an important aspect of motor control and for which the cerebellum plays a key role. Another action is to learn to associate two independent processes with each other. An air puff to the eye is perceived as unpleasant, leading to a blink reflex. If the air puff is preceded by an auditory signal, an association is formed between the two stimuli so that the auditory signal elicits a blink reflex before the air puff is presented (McCormick & Thompson, 1984 a, b; Thompson & Krupa, 1994; Jirenhed et al., 2017). This depends on processes in a specific microregion of the cerebellar C3 zone. The climbing fibers become activated by the air puff, which in turn affects the Purkinje cell, which then can lead to synaptic plastic changes in the parallel fibers activated by the auditory input (Boele et al., 2010; Thompson & Krupa, 1994; Jirenhed et al., 2017). Plasticity at the level of the dentate nucleus also contributes (McCormick & Thompson, 1984 a, b).

6.7 MODELING AND SIMULATION OF THE CEREBELLAR CIRCUITRY

The stereotypic circuitry over the entire cerebellar cortex, with specific input for each little part of the cerebellum and conserved circuitry and output structure, suggests that the computation in each specific part of cerebellum is similar. It depends essentially on the input provided to the cerebellar microcircuits, and the output effects are a function of the efferent connections. Marr (1969) and Albus (1971) each made a conceptual model of cerebellar learning, which caused Ito (1972) to suggest, based on detailed experimentation, essentially the current hypothesis of cerebellar function (Lisberger, 2021). The climbing fiber input is considered as the error signal, and the learning is conceived to occur in the parallel fiber input based on the work on the vestibulo-ocular reflex. Additional layers of understanding have been added since 1972, however (e.g., Kawato & Gomi, 1992; Kawato et al., 1987, 2021; Tokuda et al., 2017), considering the interaction between the cerebellum and the visuomotor control via the neocortex controlling biologically based hominoid robots. Their visuomotor coordination was even sufficient for the robot to successfully play ping-pong. Detailed data-driven modeling of the Purkinje cells with their extensive dendritic tree, role in plasticity, and the related cellular microcircuits have provided important insights (Zang et al., 2020; de Schutter, 1994). The detailed simulation of the integrated cerebellar machinery is currently performed in the laboratory of Egidio d'Angelo and studies are looking at how it can be used to control downstream motor centers (e.g., Bares et al., 2019).

6.8 CONCLUDING REMARKS: THE OVERALL ROLE OF THE CEREBELLUM

To summarize, lesions of the cerebellum lead to a lack of precision in the movements that are performed; although they can be initiated, the movements are generally slower and less coordinated than normal. The cerebellum receives input from all parts of the nervous system conveyed via afferents terminating as mossy fibers onto granule cells, providing very detailed information on what is going on in any circuit from the spinal cord to the neocortex. Each of the numerous modules (microzones) of the cerebellum most likely does its own thing, so there is an extensive parallel processing. This information is forwarded by numerous granule cells through the parallel fibers to the Purkinje cells. A common principle is that information is forwarded on events mediated from proprioceptors in muscles and joints or other types of receptors, which can be compared with the central commands for execution of specific movements (what is referred to as "efference copies"). For locomotion, different pathways to the cerebellum signal the flexor and extensor commands being issued, compared with information on how the movements actually proceed. The cerebellum is thus a dynamic storehouse providing information on what is happening in each corner of the nervous system.

The most important function of the cerebellum is to help correct movements when there is a dysfunction, such as when the vestibulo-ocular reflex needs to be recalibrated, if one gets new glasses, or locomotor movements need to be modified due to a wound that activates pain receptors. The error message is transmitted through neurons in the inferior olive via climbing fibers that activate Purkinje cells, which in turn can change the gain in the parallel fiber synapses onto the Purkinje cells. The climbing fiber acts through excitation of the entire dendritic tree of the Purkinje cells, which can induce synaptic plasticity in the parallel fiber synapses and thereby recalibrate the movements. Cerebellar circuits can also learn to associate two independent events with each other, such as with conditioned reflexes (Jirenhed et al., 2017).

7 COMMENTS ON WHAT WE HAVE LEARNED AND THE CHALLENGES AHEAD

In the preceding chapters of this book, I have explained how I interpret and where we, the neuroscience community, stand in terms of our understanding of how the brain can make us move. We now have a much deeper understanding of how movement is controlled than when I started as a PhD student, but there are still formidable challenges ahead.

We have discussed the evolution of the many different classes of movements. All basic features—the blueprint—of the vertebrate nervous system had already evolved at the point when lampreys diverged from the evolutionary line leading to mammals some 500 million years ago. Essential features of the vertebrate nervous system were at this time already present, such as the sensory and motor areas of the cortex/pallium, the basal ganglia with the dopamine system, the midbrain, and circuits in the tectum for detection of salient stimuli in the surrounding space. The many microcircuits in the brainstem and spinal cord were also present, but new ones have evolved to handle elements such as the control of the appendages.

From very small brains such as that of the lamprey, the number of neurons has progressively increased, and with that the potential for more sophisticated processing. For instance, the number of cortical columns in humans are orders of magnitude larger than that of the mouse. Many animals have a special niche in which their motor behavior is much more impressive than that of humans. Consider, for instance, a monkey swinging from tree to tree and a squirrel jumping from one branch to another, excelling in outstanding visuomotor control, or a passerine balancing happily on a thin branch of a tree while the wind is blowing. What is characteristic of motion control in the primate (including humans), however, is the versatility of the motor system and the possibility for endless combinations. Think about the human hand and the large numbers of configurations in which the fingers can be used, such as when one carries fragile objects to the table, writes, or performs even more skilled actions like those of a goldsmith or a pianist. The most astounding and unique human achievement through evolution is that of developing speech and the cognitive underpinning for language.

7.1 "TO MOVE OR NOT TO MOVE," A QUESTION ANSWERED BY THE BASAL GANGLIA IN CLOSE INTERACTION WITH THE CORTEX

In the introductory text of this book, I wrote that the motor infrastructure (Grillner, 2003) of preformed circuits such as those of locomotion, reaching, and oculomotor control are like the members of an orchestra responsible for the execution, while the forebrain corresponds to the conductor, determining when each motor program should be called into action. I use the term "forebrain" because I believe that the cortex and basal ganglia function together as an integrated whole—one is not understandable without the other, although they play complementary roles (figure 7.1). The following scenario would seem the most likely, based on what is now known (at least to me).

The input to the forebrain is via the thalamus, which distributes information from all the senses, the midbrain, and the microcircuits in the brainstem and spinal cord to the various areas of the cortex and striatum. In mammals, the neocortex processes information from their surroundings in a very elaborate way. Most of our movements depend on interactions with objects around us and require a visual analysis of the precise conditions required to grasp an object or perform action. When in contact, an interpretation of the haptic information from the hand (chapter 5) is needed to maintain the grip and not to lose contact. During locomotion, we similarly orient our movements in relation to the environment based mostly on visual information. Although some processing may take place at the midbrain level, the task of the cortex in this context is to interpret the surrounding world and forward this information to the frontal lobe and to the striatum for possible action and reaction.

Each part of the neocortex projects to its specific target area within the striatum (Foster et al., 2021). For instance, the motor areas in the frontal lobe communicate with specific motor areas in the forelimb, hindlimb, or orofacial areas in the dorsolateral striatum, while the associate and limbic cortices target other areas in the dorsomedial or ventral striatum. Thalamo-striatal afferents contribute to around 40 percent on the input to the striatum regarding brainstem and cerebellar activity (chapter 4).

For all everyday motor activities, such as initiating saccadic eye movements, locomotion, or reaching for an object, the decision process is most likely based on cortical or thalamic inputs that target the relevant parts of the striatum, such as the forelimb area. If the input to the striatum is strong enough (perhaps amplified by dopamine action), and if there are no competing actions, it may lead to an activation of the striatal projection neurons, which in turn inhibit the appropriate subset of substantia nigra pars reticulata (SNr) neurons. This leads to disinhibition of downstream motor centers and allows an action to be initiated (chapters 4 and 5). The decision thus appears to depend on an initiation process that originates in the cortex, thalamus, or other structures but is completed within the striatum. The cortical input to the striatum is in part conveyed via the pyramidal neuron (PT) in layer 5, which is all divided into subgroups depending on target areas and which types of movement they elicit. One would expect that the PT neurons that activate a given group of striatal circuits that promotes a certain action of a given muscle group also provide direct excitation of the same group of neurons at the brainstem level. This would suggest that the combined action of the forebrain in the control of movement is to allow movement through the disinhibition via the basal ganglia circuit and aided by direct excitation from PT neurons—a push-pull arrangement or a dance in which the cortex and basal ganglia complement each other.

In this process, dopamine plays an important role. The initiation of movement is often preceded by a burst in dopamine neurons, perhaps prompted by a salient event that will affect the excitability within the striatum. That dopamine is central in the decision process is clear from the difficulties to initiate movements that occur in Parkinson's disease and dopamine-depleted animals, and conversely the hyperkinesia that result from excessive dopamine levels that can occur with factors such as L-DOPA medication.

7.2 THE MAJOR ORGANIZATIONAL BUILDING BLOCKS OF MOTION

Figure 7.1 summarizes in a cartoonlike fashion how one can view the motor system in a very schematic way. Instinctive behavior (chapters 2 and 3), such as escape (freezing and aggression), foraging (food and fluid intake), and reproductive behaviors (pairing and maternal behavior), can be triggered from the hypothalamus and channeled through the compartments of the periaqueductal gray (PAG) before activating the downstream motor circuits. In the basal ganglia, the SNr has an inhibitory effect on the PAG, all the downstream microcircuits in the midbrain–brainstem, and the motor centers that control the spinal cord via descending pathways. In an emergency, such as escape, the excitation may override the SNr inhibition unless the basal ganglia/substantia nigra pars compacta (SNc) circuit has not had the time to lift the inhibition.

7.3 THE ROLE OF THE CEREBELLUM: THE PERFECTION OF COORDINATION

Not included in figure 7.1 is the cerebellum, since it plays a different role from that of the circuits described in the previous section, which are concerned with

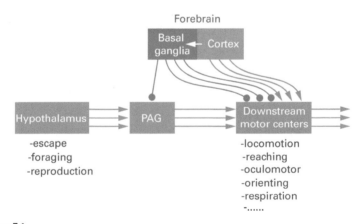

Figure 7.1
Organization of the major building blocks of the vertebrate motor system. Color-coding: blue is inhibitory and red excitatory.

the initiation and execution of different movements. After lesions have been made in the cerebellum, movements can still be initiated, but the quality is degraded (chapter 6). The cerebellum receives information from all parts of the nervous system and serves as a dynamic storehouse of what happens in each part of the nervous system, from the caudal tip of the spinal cord to any part of the neocortex. The cerebellum can through its efferent connections fine-tune the activity of most motor circuits through a form of motor learning. Error signals from the circuits that have a suboptimal function are mediated via the inferior olive to the cerebellum, whether we are concerned with vestibulo-ocular reflexes (retinal slip), locomotion, or any other motor pattern (including speech). The cerebellum is crucial for the overall performance of the motor system, as it contributes to the perfection and fine-tuning of the movements.

To summarize, the cortex and the basal ganglia in close interaction determine which specific movements should be recruited at any point of time, while the execution is mostly controlled by circuits ranging from the midbrain to the spinal cord level. The fine-tuning of each type of movement is one major task of the cerebellum.

7.4 SOME CHALLENGES AHEAD

7.4.1 Visuomotor Coordination

In our everyday lives, humans interact with objects around us—grab a cup of coffee, press the handle of a door, or unlock the door with a key. These are simple tasks, but they all usually depend on vision to identify the object we need to interact with, positioning the hands or feet appropriately to perform the motor action. We know that there is a need for visuomotor transformation and part of the dorsal stream in the cortex is involved, and some neurons may respond to a preferred preshaping of the hand before actually grasping. However, the details of these visuomotor transformations are yet to be understood.

7.4.2 Combining Different Forms of Motion into an Integrated, Harmonious Whole

A ballet dancer onstage or a tennis player on the court combines a variety of movements dynamically, and one motor pattern merges into another in a seamless way. We may be able to account for each individual part of the motor pattern, but how the brain is able to combine them into a graceful whole is still a mystery. The same applies to a cheetah trying to hunt down prey. We know that the basal ganglia contribute since Parkinsonian patients lose the ability to combine movements and are forced to perform one motor pattern at a time (chapter 4). How the basal ganglia is able to handle this remains to be discovered.

7.4.3 Motor Learning

Our understanding of how learning in the motor system occurs is limited. In the cerebellum, we know that the vestibulo-ocular reflex and other networks can be recalibrated through the interaction between the climbing and parallel fibers, and two independent processes can be associated, as in conditioned reflexes. Changes in synaptic strength can be induced in many parts of the nervous system through long-term potentiation (LTP) and long-term depression (LTD), building blocks for motor learning. In the basal ganglia, in interaction with the dopamine system, reinforcement learning is most likely an important principle. The next level of understanding of how and where the learned motor microcircuits are stored and recruited is rather foggy, although we know that certain structures of the nervous system need to operate in order to remember or recall certain memories, but not how this occurs.

7.4.4 Human Speech and Language

The most spectacular human motor repertoire is connected to language and speech motor control. We know which areas of the human brain that are involved when considering different aspects of language: Wernicke's area for understanding what is said, and Broca's area in the left hemisphere for the motor aspect of the spoken word, with the prosody contributing on the corresponding area of the right side of the brain (chapter 5). Many parts of the cortex are involved in speech, but our knowledge is for the most part limited to which areas of the cortex are involved, and with only some knowledge of neurons in the frontal lobe that are specific to different vowels. Each vowel is produced by the tongue having a specific position in the oral cavity. A fair bit of knowledge is available about the biomechanics of speech, including how we can produce the phonemes in rapid succession. The essential question of the neural mechanisms used for learning to talk are actually not understood and at this point are only described phenomenologically, with no understanding of the intrinsic microcircuit mechanisms. Imaging techniques with ultrahigh resolution regarding both temporal and spatial resolution can further advance our knowledge, but still only on a mesoscopic level.

Over a period of several years, the language ability develops, from the babbling of the baby to the formation of individual words and then of sentences in the infant to finally the ability to speak fluently in one, two, or several languages. A native English speaker can have a vocabulary ranging from 10,000 to 65,000 words. However, we have very little knowledge of how this all happens, how words are stored, how they readily are recruited when talking, and how we can match the corresponding words of different languages to each other. Furthermore, how we easily apply the grammatical rules of the specific language that we are currently speaking is not understood.

These questions cannot be addressed easily in animal models. The most relevant example may be birdsong. In most cases, young male birds hear the song while in the nest. Later, when their hormones indicate that it is time to start singing, they begin to attempt to do so, and gradually improve their ability. Most birds have some form of innate template and readily learn the song of their own species. However, by exposing young birds to another type of song, they can within limits learn that new song. Some species, such as the mockingbird and the parrot, can pick up many sounds from the surrounding world, including human words, and they learn by imitation. Nevertheless, there is a quantum leap in the complexity between birdsong and language, although in both cases there is a phase of learning by imitation—hearing a sound and then being able to perform a coordinated motor act that produces the sound just heard. The message conveyed by the birdsong is twofold: to scare away other males and to attract females. Human language, on the other hand, has more sophisticated content. In addition, the birds, like other vertebrates, have a series of innate sounds used for communication, such as warning calls.

7.4.5   One Ultimate Question Is Concerned with the Aspect of Volition in the Control of Movement

We all have the ability on the spur of the moment to initiate a movement of the index finger or a sidestep without any external reason—we just know that we somehow like to do that, and we have no idea of how this happens. We noted in chapter 5 that when the supplementary motor cortex was stimulated, the resulting arm movement was perceived by the patient as if he or she chose to move the arm. In contrast, if the motor cortex was stimulated, it was perceived as an unexpected jerk. As we have seen in this text, we have a fair understanding of the neural infrastructure that underlies the execution of movement. However, the neurobiology of the neural mechanisms behind what we perceive as voluntary is still unknown, and I anticipate that these mechanisms exist in other mammals as well.

To conclude, the brain offers a number of interesting, fundamental, and challenging problems. Some are close to being solved, and others have remained more remote. As neuroscientists, we do not run the risk of being out of work, and we can enjoy the marked progress of the preceding decades and look forward to an exciting future for neuroscience.

References

Abecassis, Z. A., Berceau, B. L., Win, P. H., Pamukcu, A., Cherian, S., Hernandez, V. M., Chon, U., Lim, B. K., Kim, Y., Justice, N. J., Awatramani, R., Hooks, B. M., Gerfen, C. R., Boca, S. M., & Chan, C. S. (2020). Npas1(+)-Nkx2.1(+) Neurons are an integral part of the cortico-pallido-cortical loop. *J. Neurosci., 40*, 743–768.

Adhikari, A., Lerner, T. N., Finkelstein, J., Pak, S., Jennings, J. H., Davidson, T. J., Ferenczi, E., Gunaydin, L. A., Mirzabekov, J. J., Ye, L., Kim, S. Y., Lei, A., & Deisseroth, K. (2015). Basomedial amygdala mediates top-down control of anxiety and fear. *Nature, 527*, 179–185.

Adolphs, R., & Anderson, D. J. (2018). *The neuroscience of emotion—A new synthesis.* Princeton University Press.

Albus, J. S. (1971). A theory of cerebellar function. *Math Biosci., 10*, 25–61.

Alford, S., & Williams, T. L. (1989). Endogenous activation of glycine and NMDA receptors in lamprey spinal cord during fictive locomotion. *J. Neurosci., 9*, 2792–2800.

Alstermark, B., & Isa, T. (2012). Circuits for skilled reaching and grasping. *Annu. Rev. Neurosci., 35*, 559–578.

Alstermark, B., Isa, T., Pettersson, L. G., & Sasaki, S. (2007). The C3-C4 propriospinal system in the cat and monkey: A spinal pre-motoneuronal centre for voluntary motor control. *Acta. Physiol. (Oxf.), 189*, 123–140.

Amemori, S., Graybiel, A. M., & Amemori, K. I. (2021). Causal evidence for induction of pessimistic decision-making in primates by the network of frontal cortex and striosomes. *Front. Neurosci., 15*, 649167.

An, X., Matho, K., Li, Y., Mohan, H., Xu, X. H., Whishaw, I. Q., Kepecs, A., & Huang, J. (2022). A cortical circuit for orchestrating oromanual food manipulationcoordinates food handling and manipulation. *BioRxiv.* doi:10.1101/2022.12.03.518964.

Anderson, T. M., Garcia, A. J., 3rd, Baertsch, N. A., Pollak, J., Bloom, J. C., Wei, A. D., Rai, K. G., & Ramirez, J. M. (2016). A novel excitatory network for the control of breathing. *Nature, 536*, 76–80.

Andersson, B. (1953). The effect of injections of hypertonic NaCl-solutions into different parts of the hypothalamus of goats. *Acta Physiol. Scand.*, *28*, 188–201.

Andersson, O., Forssberg, H., Grillner, S., & Wallen, P. (1981). Peripheral feedback mechanisms acting on the central pattern generators for locomotion in fish and cat. *Can. J. Physiol. Pharmacol.*, *59*, 713–726.

Anis, E., Xie, A., Brundin, L., & Brundin, P. (2021). Digesting recent findings: Gut alpha-synuclein, microbiome changes in Parkinson's disease. *Trends. Endocrinol. Metab.*, *33*(2), 147–157.

Apps, R., Hawkes, R., Aoki, S., Bengtsson, F., Brown, A. M., Chen, G., Ebner, T. J., Isope, P., Jörntell, H., Lackey, E. P., Lawrenson, C., Lumb, B., Schonewille, M., Sillitoe, R. V., Spaeth, L., Sugihara, I., Valera, A., Voogd, J., Wylie, D. R., & Ruigrok, T. J. H. (2018). Cerebellar modules and their role as operational cerebellar processing units: A consensus paper [corrected]. *Cerebellum, 17*, 654–682.

Arber, S., & Costa, R. M. (2018). Connecting neuronal circuits for movement. *Science, 360*, 1403–1404.

Arshavsky, Y. I., Berkinblit, M. B., Fukson, O. I., Gelfand, I. M., & Orlovsky, G. N. (1972). Recordings of neurones of the dorsal spinocerebellar tract during evoked locomotion. *Brain Res.*, *43*, 272–275.

Arshavsky, Y. I., Gelfand, I. M., Orlovsky, G. N., & Pavlova, G. A. (1978a). Messages conveyed by descending tracts during scratching in the cat. I. Activity of vestibulospinal neurons. *Brain Res.*, *159*, 99–110.

Arshavsky, Y. I., Orlovsky, G. N., Pavlova, G. A., & Perret, C. (1978b). Messages conveyed by descending tracts during scratching in the cat. II. Activity of rubrospinal neurons. *Brain Res.*, *159*, 111–123.

Arshavsky Yu, I., Gelfand, I. M., Orlovsky, G. N., Pavlova, G. A., & Popova, L. B. (1984). Origin of signals conveyed by the ventral spino-cerebellar tract and spino-reticulo-cerebellar pathway. *Exp. Brain. Res.*, *54*, 426–431.

Ashhad, S., & Feldman, J. L. (2020). Emergent elements of inspiratory rhythmogenesis: Network synchronization and synchrony propagation. *Neuron, 106*, 482–497 e484.

Assous, M. (2021). Striatal cholinergic transmission. Focus on nicotinic receptors' influence in striatal circuits. *Eur. J. Neurosci.*, *53*, 2421–2442.

Azim, E., Jiang, J., Alstermark, B., & Jessell, T. M. (2014). Skilled reaching relies on a V2a pro-priospinal internal copy circuit. *Nature, 508*, 357–363.

Ballester, B. R., Maier, M., Duff, A., Cameirao, M., Bermudez, S., Duarte, E., Cuxart, A., Rodri-guez, S., San Segundo Mozo, R. M., & Verschure, P. (2019). A critical time window for recovery extends beyond one-year post-stroke. *J. Neurophysiol.*, *122*, 350–357.

Bareš, M., Apps, R., Avanzino, L., Breska, A., D'Angelo, E., Filip, P., Gerwig, M., Ivry, R. B., Lawrenson, C. L., Louis, E. D., Lusk, N. A., Manto, M., Meck, W. H., Mitoma, H., & Petter,

E. A. (2019). Consensus paper: Decoding the contributions of the cerebellum as a time machine. From neurons to clinical applications. *Cerebellum, 18*, 266–286.

Beloozerova, I. N., & Sirota, M. G. (1986). Activity of neurons of the motor-sensory cortex of the cat during natural locomotion while stepping over obstacles. *Neirofiziologiia, 18*, 546–549.

Beltramo, R., & Scanziani, M. (2019). A collicular visual cortex: Neocortical space for an ancient midbrain visual structure. *Science, 363*, 64–69.

Benabid, A. L., Chabardes, S., Torres, N., Piallat, B., Krack, P., Fraix, V., & Pollak, P. (2009). Functional neurosurgery for movement disorders: A historical perspective. *Prog. Brain. Res., 175*, 379–391.

Benavidez, N. L., Bienkowski, M. S., Zhu, M., Garcia, L. H., Fayzullina, M., Gao, L., Bowman, I., Gou, L., Khanjani, N., Cotter, K. R., Korobkova, L., Becerra, M., Cao, C., Song, M. Y., Zhang, B., Yamashita, S., Tugangui, A. J., Zingg, B., Rose, K., Lo, D., Foster, N. N., Boesen, T., Mun, H. S., Aquino, S., Wickersham, I. R., Ascoli, G. A., Hintiryan, H., & Dong, H. W. (2021). Organization of the inputs and outputs of the mouse superior colliculus. *Nat. Commun., 12*, 4004.

Benazzouz, A., Gross, C., Feger, J., Boraud, T., & Bioulac, B. (1993). Reversal of rigidity and improvement in motor performance by subthalamic high-frequency stimulation in MPTP-treated monkeys. *Eur. J. Neurosci., 5*, 382–389.

Bergman, H., Wichmann, T., & DeLong, M. R. (1990). Reversal of experimental parkinsonism by lesions of the subthalamic nucleus. *Science, 249*, 1436–1438.

Bergman, H., Wichmann, T., Karmon, B., & DeLong, M. R. (1994). The primate subthalamic nucleus. II. Neuronal activity in the MPTP model of parkinsonism. *J. Neurophysiol., 72*, 507–520.

Bernstein, N. (1967). *The coordination and regulation of movements*. Pergamon.

Björkman, A., Weibull, A., Rosen, B., Svensson, J., & Lundborg, G. (2009). Rapid cortical reorganisation and improved sensitivity of the hand following cutaneous anaesthesia of the forearm. *Eur. J. Neurosci., 29*, 837–844.

Bjursten, L. M., Norrsell, K., & Norrsell, U. (1976). Behavioural repertory of cats without cerebral cortex from infancy. *Exp. Brain. Res., 25*, 115–130.

Boele, H. J., Koekkoek, S. K., & De Zeeuw, C. I. (2010). Cerebellar and extracerebellar involvement in mouse eyeblink conditioning: The ACDC model. *Front. Cell. Neurosci., 3*, 19.

Bolam, J. P., & Pissadaki, E. K. (2012). Living on the edge with too many mouths to feed: Why dopamine neurons die. *Mov. Disord., 27*, 1478–1483.

Bolam, J. P., Wainer, B. H., & Smith, A. D. (1984). Characterization of cholinergic neurons in the rat neostriatum. A combination of choline acetyltransferase immunocytochemistry, Golgi-impregnation and electron microscopy. *Neuroscience, 12*, 711–718.

Bostan, A. C., & Strick, P. L. (2018). The basal ganglia and the cerebellum: Nodes in an integrated network. *Nat. Rev. Neurosci., 19*, 338–350.

Boyes, J., & Bolam, J. P. (2007). Localization of GABA receptors in the basal ganglia. *Prog. Brain Res.*, *160*, 229–243.

Braak, H., Del Tredici, K., Rub, U., de Vos, R. A., Jansen Steur, E. N., & Braak, E. (2003). Staging of brain pathology related to sporadic Parkinson's disease. *Neurobiol. Aging*, *24*, 197–211.

Brinkman, C. (1984). Supplementary motor area of the monkey's cerebral cortex: Short- and long-term deficits after unilateral ablation and the effects of subsequent callosal section. *J. Neurosci.*, *4*, 918–929.

Brodin, L., Buchanan, J. T., Hokfelt, T., Grillner, S., & Verhofstad, A. A. (1986). A spinal projection of 5-hydroxytryptamine neurons in the lamprey brainstem; evidence from combined retrograde tracing and immunohistochemistry. *Neurosci. Lett.*, *67*, 53–57.

Brodin, L., Hökfelt, T., Grillner, S., & Panula, P. (1990). Distribution of histaminergic neurons in the brain of the lamprey Lampetra fluviatilis as revealed by histamine-immunohistochemistry. *J. Comp. Neurol.*, *292*, 435–442.

Bruce, C. J., Goldberg, M. E., Bushnell, M. C., & Stanton, G. B. (1985). Primate frontal eye fields. II. Physiological and anatomical correlates of electrically evoked eye movements. *J. Neurophysiol.*, *54*, 714–734.

Bruce, N. J., Narzi, D., Trpevski, D., van Keulen, S. C., Nair, A. G., Rothlisberger, U., Wade, R. C., Carloni, P., & Hellgren Kotaleski, J. (2019). Regulation of adenylyl cyclase 5 in striatal neurons confers the ability to detect coincident neuromodulatory signals. *PLoS Comput. Biol.*, *15(10)*, e1007382.

Buchanan, J. T., & Grillner, S. (1987). Newly identified "glutamate interneurons" and their role in locomotion in the lamprey spinal cord. *Science, 236*, 312–314.

Buonomano, D. V., & Merzenich, M. M. (1998). Cortical plasticity: From synapses to maps. *Annu. Rev. Neurosci.*, *21*, 149–186.

Burke, D. A., Rotstein, H. G., & Alvarez, V. A. (2017). Striatal local circuitry: A new framework for lateral inhibition. *Neuron, 96*, 267–284.

Caboche, J., Vanhoutte, P., Boussicault, L., & Betuing, S. (2017). Cellular and molecular mechanisms of neuronal dysfunction in Huntington's disease. In H. Steiner & K. Y. Tseng (Eds.), *Handbook of basal ganglia structure and function* (pp. 889–906). Academic Press.

Caggiano, V., Leiras, R., Goni-Erro, H., Masini, D., Bellardita, C., Bouvier, J., Caldeira, V., Fisone, G., & Kiehn, O. (2018). Midbrain circuits that set locomotor speed and gait selection. *Nature, 553*, 455–460.

Caligiore, D., Pezzulo, G., Baldassarre, G., Bostan, A. C., Strick, P. L., Doya, K., Helmich, R. C., Dirkx, M., Houk, J., Jörntell, H., Lago-Rodriguez, A., Galea, J. M., Miall, R. C., Popa, T., Kishore, A., Verschure, P. F., Zucca, R., & Herreros, I. (2017). Consensus paper: Towards a systems-level view of cerebellar function: The interplay between cerebellum, basal ganglia, and cortex. *Cerebellum, 16*, 203–229.

Cangiano, L., & Grillner, S. (2005). Mechanisms of rhythm generation in a spinal locomotor network deprived of crossed connections: The lamprey hemicord. *J. Neurosci.*, *25*, 923–935.

Capantini, L., von Twickel, A., Robertson, B., & Grillner, S. (2017). The pretectal connectome in lamprey. *J. Comp. Neurol.*, *525*, 753–772.

Capelli, P., Pivetta, C., Soledad Esposito, M., & Arber, S. (2017). Locomotor speed control circuits in the caudal brainstem. *Nature, 551*, 373–377.

Carlsson, A. (1964). Evidence for a role of dopamine in extrapyramidal functions. *Acta. Neuroveg. (Wien)*, *26*, 484–493.

Carlsson, A. (2001). A paradigm shift in brain research. *Science, 294*, 1021–1024.

Carr, C. E., & Konishi, M. (1990). A circuit for detection of interaural time differences in the brain stem of the barn owl. *J. Neurosci.*, *10*, 3227–3246.

Cenci, M. A. (2017). Molecular mechanisms of l-DOPA-induced dyskinesia. In H. Steiner & K. Y. Tseng (Eds.), *Handbook of basal ganglia structure and function* (pp. 857–871). Academic Press.

Chen, C. H., Fremont, R., Arteaga-Bracho, E. E., & Khodakhah, K. (2014). Short latency cerebellar modulation of the basal ganglia. *Nat. Neurosci.*, *17*, 1767–1775.

Cheong, R. Y., Baldo, B., Sajjad, M. U., Kirik, D., & Petersen, A. (2021). Effects of mutant huntingtin inactivation on Huntington disease–related behaviours in the BACHD mouse model. *Neuropathol. Appl. Neurobiol.*, *47*, 564–578.

Cherng, B. W., Islam, T., Torigoe, M., Tsuboi, T., & Okamoto, H. (2020). The dorsal lateral habenula-interpeduncular nucleus pathway is essential for left-right-dependent decision making in zebrafish. *Cell Rep., 32*, 108143. https://doi.org/10.1016/j.celrep.2020.108143

Cinelli, E., Mutolo, D., Contini, M., Pantaleo, T., & Bongianni, F. (2016). Inhibitory control of ascending glutamatergic projections to the lamprey respiratory rhythm generator. *Neuroscience, 326*, 126–140.

Cinelli, E., Robertson, B., Mutolo, D., Grillner, S., Pantaleo, T., & Bongianni, F. (2013). Neuronal mechanisms of respiratory pattern generation are evolutionary conserved. *J. Neurosci.*, *33*, 9104–9112.

Cisek, P., & Kalaska, J. F. (2005). Neural correlates of reaching decisions in dorsal premotor cortex: Specification of multiple direction choices and final selection of action. *Neuron., 45*, 801–814.

Clark, F. J., & von Euler, C. (1972). On the regulation of depth and rate of breathing. *J. Physiol.*, *222*, 267–295.

Comoli, E., Coizet, V., Boyes, J., Bolam, J. P., Canteras, N. S., Quirk, R. H., Overton, P. G., & Redgrave, P. (2003). A direct projection from superior colliculus to substantia nigra for detecting salient visual events. *Nat. Neurosci.*, *6*, 974–980.

Comoli, E., Ribeiro-Barbosa, E. R., & Canteras, N. S. (2000). Afferent connections of the dorsal premammillary nucleus. *J. Comp. Neurol.*, *423*, 83–98.

Condamine, S., Lavoie, R., Verdier, D., & Kolta, A. (2018). Functional rhythmogenic domains defined by astrocytic networks in the trigeminal main sensory nucleus. *Glia, 66*, 311–326.

Cooper, B., & McPeek, R. M. (2021). Role of the superior colliculus in guiding movements not made by the eyes. *Annu. Rev. Vis. Sci., 7*, 279–300.

Crittenden, J. R., & Graybiel, A. M., (2011). Basal ganglia disorders associated with imbalances in the striatal striosome and matrix compartments. *Front. Neuroanat., 5*, 59.

Crittenden, J. R., Lacey, C. J., Weng, F. J., Garrison, C. E., Gibson, D. J., Lin, Y., & Graybiel, A. M. (2017). Striatal cholinergic interneurons modulate spike-timing in striosomes and matrix by an amphetamine-sensitive mechanism. *Front. Neuroanat., 11*, 20.

Cui, Q., Du, X., Chang, I. Y. M., Pamukcu, A., Lilascharoen, V., Berceau, B. L., Garcia, D., Hong, D., Chon, U., Narayanan, A., Kim, Y., Lim, B. K., & Chan, C. S. (2021). Striatal direct pathway targets Npas1(+) pallidal neurons. *J. Neurosci., 41*, 3966–3987.

Daie, K., Svoboda, K., & Druckmann, S. (2021). Targeted photostimulation uncovers circuit motifs supporting short-term memory. *Nat. Neurosci., 24*, 259–265.

Darmohray, D. M., Jacobs, J. R., Marques, H. G., & Carey, M. R. (2019). Spatial and temporal locomotor learning in mouse cerebellum. *Neuron, 102*, 217–231 e214.

Darwin, C. (1872). *The expression of the emotions in man and animals*. John Murray.

Dasen, J. S., & Jessell, T. M. (2009). Hox networks and the origins of motor neuron diversity. *Curr. Top. Dev. Biol., 88*, 169–200.

da Silva, J. A., Tecuapetla, F., Paixao, V., & Costa, R. M. (2018). Dopamine neuron activity before action initiation gates and invigorates future movements. *Nature, 554*, 244–248.

Dean, P., Redgrave, P., & Westby, G. W. (1989). Event or emergency? Two response systems in the mammalian superior colliculus. *Trends Neurosci., 12*, 137–147.

DeFelipe, J., Alonso-Nanclares, L., & Arellano, J. I. (2002). Microstructure of the neocortex: Comparative aspects. *J. Neurocytol., 31*, 299–316.

Dellow, P. G., & Lund, J. P. (1971). Evidence for central timing of rhythmical mastication. *J. Physiol., 215*, 1–13.

Del Negro, C. A., Funk, G. D., & Feldman, J. L. (2018). Breathing matters. *Nat. Rev. Neurosci., 19*, 351–367.

DeLong, M. R., Crutcher, M. D., & Georgopoulos, A. P. (1985). Primate globus pallidus and subthalamic nucleus: Functional organization. *J. Neurophysiol., 53*, 530–543.

de Manzano, O., Kuckelkorn, K. L., Ström, K., & Ullén, F. (2020). Action-perception coupling and near transfer: Listening to melodies after piano practice triggers sequence-specific representations in the auditory-motor network. *Cereb. Cortex., 30*, 5193–5203.

de Schutter, E. (1994). Modelling the cerebellar Purkinje cell: Experiments in computo. *Prog. Brain Res., 102*, 427–441.

De Zeeuw, C. I. (2021). Bidirectional learning in upbound and downbound microzones of the cerebellum. *Nat. Rev. Neurosci.*, *22*, 92–110.

Dhawale, A. K., Smith, M. A., & Olveczky, B. P. (2017). The role of variability in motor learning. *Annu. Rev. Neurosci.*, *40*, 479–498.

Dhawale, A. K., Wolff, S. B. E., Ko, R., & Olveczky, B. P. (2021). The basal ganglia control the detailed kinematics of learned motor skills. *Nat. Neurosci.*, *24*, 1256–1269.

Diederich, N. J., Surmeier, D. J., Uchihara, T., Grillner, S., & Goetz, C. G. (2019). Parkinson's disease: Is it a consequence of human brain evolution? *Mov. Disord.*, *34*, 453–459.

Diederich, N. J., Uchihara, T., Grillner, S., & Goetz, C. G. (2020). The evolution-driven signature of Parkinson's disease. *Trends Neurosci.*, *43*, 475–492.

DiMarco, A. F., Romaniuk, J. R., von Euler, C., & Yamamoto, Y. (1983). Immediate changes in ventilation and respiratory pattern associated with onset and cessation of locomotion in the cat. *J. Physiol.*, *343*, 1–16.

Ding, J. B., Guzman, J. N., Peterson, J. D., Goldberg, J. A., & Surmeier, D. J. (2010). Thalamic gating of corticostriatal signaling by cholinergic interneurons. *Neuron*, *67*, 294–307.

di Pellegrino, G., Fadiga, L., Fogassi, L., Gallese, V., & Rizzolatti, G. (1992). Understanding motor events: A neurophysiological study. *Exp. Brain. Res.*, *91*, 176–180.

Di Prisco, G. V., Wallén, P., Grillner, S. (1990). Synaptic effects of intraspinal stretch receptor neurons mediating movement-related feedback during locomotion. *Brain. Res.*, *530*, 161–166.

Doig, N. M., Moss, J., & Bolam, J. P. (2010). Cortical and thalamic innervation of direct and indirect pathway medium-sized spiny neurons in mouse striatum. *J. Neurosci.*, *30*, 14610–14618.

Dominici, N., Ivanenko, Y. P., Cappellini, G., d'Avella, A., Mondi, V., Cicchese, M., Fabiano, A., Silei, T., Di Paolo, A., Giannini, C., Poppele, R. E., & Lacquaniti, F. (2011). Locomotor primitives in newborn babies and their development. *Science*, *334*, 997–999.

Dorst, M. C., Tokarska, A., Zhou, M., Lee, K., Stagkourakis, S., Broberger, C., Masmanidis, S., & Silberberg, G. (2020). Polysynaptic inhibition between striatal cholinergic interneurons shapes their network activity patterns in a dopamine-dependent manner. *Nat. Commun.*, *11*, 5113.

Dougherty, K. J., Zagoraiou, L., Satoh, D., Rozani, I., Doobar, S., Arber, S., Jessell, T. M., & Kiehn, O. (2013). Locomotor rhythm generation linked to the output of spinal shox2 excitatory interneurons. *Neuron*, *80*, 920–933.

Downey, J. E., Weiss, J. M., Muelling, K., Venkatraman, A., Valois, J. S., Hebert, M., Bagnell, J. A., Schwartz, A. B., & Collinger, J. L. (2016). Blending of brain-machine interface and vision-guided autonomous robotics improves neuroprosthetic arm performance during grasping. *J. Neuroeng. Rehabil.*, *13*, 28.

Drew, T., Andujar, J. E., Lajoie, K., & Yakovenko, S. (2008). Cortical mechanisms involved in visuomotor coordination during precision walking. *Brain Res. Rev.*, *57*, 199–211.

Du, K., Wu, Y. W., Lindroos, R., Liu, Y., Rozsa, B., Katona, G., Ding, J. B., & Kotaleski, J. H. (2017). Cell-type-specific inhibition of the dendritic plateau potential in striatal spiny projection neurons. *Proc. Natl. Acad. Sci. USA, 114*, E7612–E7621.

Dubbeldam, J. L., & den Boer-Visser, A. M. (2002). The central mesencephalic grey in birds: Nucleus intercollicularis and substantia grisea centralis. *Brain Res. Bull., 57*, 349–352.

Duval, C., Panisset, M., Strafella, A. P., & Sadikot, A. F. (2006). The impact of ventrolateral thalamotomy on tremor and voluntary motor behavior in patients with Parkinson's disease. *Exp. Brain Res., 170*, 160–171.

Eccles, J. C., Ito, M., & Szentágothai, J. (1967). *The cerebellum as a neuronal machine.* Springer.

Economo, M. N., Viswanathan, S., Tasic, B., Bas, E., Winnubst, J., Menon, V., Graybuck, L. T., Nguyen, T. N., Smith, K. A., Yao, Z., Wang, L., Gerfen, C. R., Chandrashekar, J., Zeng, H., Looger, L. L., & Svoboda, K. (2018). Distinct descending motor cortex pathways and their roles in movement. *Nature, 563*, 79–84.

Ekeberg, O., & Grillner, S. (1999). Simulations of neuromuscular control in lamprey swimming. *Philos. Trans. R. Soc. Lond. B. Biol. Sci., 354*, 895–902.

Ekerot, C. F., & Jörntell, H. (2008). Synaptic integration in cerebellar granule cells. *Cerebellum, 7*, 539–541.

Ekerot, C. F., & Kano, M. (1985). Long-term depression of parallel fibre synapses following stimulation of climbing fibres. *Brain Res., 342*, 357–360.

Ekerot, C. F., & Oscarsson, O. (1981). Prolonged depolarization elicited in Purkinje cell dendrites by climbing fibre impulses in the cat. *J. Physiol., 318*, 207–221.

Ekerot, C. F., Oscarsson, O., & Schouenborg, J. (1987). Stimulation of cat cutaneous nociceptive C fibres causing tonic and synchronous activity in climbing fibres. *J. Physiol., 386*, 539–546.

Ekman, P., Friesen, W. V., O'Sullivan, M., Chan, A., Diacoyanni-Tarlatzis, I., Heider, K., Krause, R., LeCompte, W. A., Pitcairn, T., Ricci-Bitti, P. E., Scherer, K., Tomita, M., & Tzavaras, A. (1987). Universals and cultural differences in the judgments of facial expressions of emotion. *J. Pers. Soc. Psychol., 53*, 712–717.

Ekstrand, M. I., Terzioglu, M., Galter, D., Zhu, S., Hofstetter, C., Lindqvist, E., Thams, S., Bergstrand, A., Hansson, F. S., Trifunovic, A., Hoffer, B., Cullheim, S., Mohammed, A. H., Olson, L., & Larsson, N. G. (2007). Progressive parkinsonism in mice with respiratory-chain-deficient dopamine neurons. *Proc. Natl. Acad. Sci. USA, 104*, 1325–1330.

Ellender, T. J., Harwood, J., Kosillo, P., Capogna, M., & Bolam, J. P. (2013). Heterogeneous properties of central lateral and parafascicular thalamic synapses in the striatum. *J. Physiol., 591*, 257–272.

El Manira, A., Pombal, M. A., & Grillner, S. (1997). Diencephalic projection to reticulospinal neurons involved in the initiation of locomotion in adult lampreys Lampetra fluviatilis. *J. Comp. Neurol., 389*, 603–616.

El Manira, A., Tegner, J., & Grillner, S. (1994). Calcium-dependent potassium channels play a critical role for burst termination in the locomotor network in lamprey. *J. Neurophysiol., 72*, 1852–1861.

Engelhard, B., Finkelstein, J., Cox, J., Venkatraman, A., Valois, J. S., Hebert, M., Bagnell, J. A., Schwartz, A. B., & Collinger, J. L. (2019). Specialized coding of sensory, motor and cognitive variables in VTA dopamine neurons. *Nature, 570*, 509–513.

Esposito, M. S., Capelli, P., & Arber, S. (2014). Brainstem nucleus MdV mediates skilled forelimb motor tasks. *Nature, 508*, 351–356.

Faber, D. S., Fetcho, J. R., & Korn, H. (1989). Neuronal networks underlying the escape response in goldfish. General implications for motor control. *Ann. NY Acad. Sci., 563*, 11–33.

Falkner, A. L., Dollar, P., Perona, P., Anderson, D. J., & Lin, D. (2014). Decoding ventromedial hypothalamic neural activity during male mouse aggression. *J. Neurosci., 34*, 5971–5984.

Falkner, A. L., Wei, D., Song, A., Watsek, L. W., Chen, I., Chen, P., Feng, J. E., & Lin, D. (2020). Hierarchical representations of aggression in a hypothalamic-midbrain circuit. *Neuron, 106*, 637–648 e636.

Faull, O. K., Subramanian, H. H., Ezra, M., & Pattinson, K. T. S. (2019). The midbrain periaqueductal gray as an integrative and interoceptive neural structure for breathing. *Neurosci. Biobehav. Rev., 98*, 135–144.

Ferreira-Pinto, M. J., Kanodia, H., Falasconi, A., Sigrist, M., Esposito, M. S., & Arber, S. (2021). Functional diversity for body actions in the mesencephalic locomotor region. *Cell, 184*, 4564–4578 e4518.

Fidelin, K., & Arber, S. (2022). *Anatomical and functional organization of red nucleus circuits*. Paper presented at FENS (Paris).

Filipovic, M., Ketzef, M., Reig, R., Aertsen, A., Silberberg, G., & Kumar, A. (2019). Direct pathway neurons in mouse dorsolateral striatum in vivo receive stronger synaptic input than indirect pathway neurons. *J. Neurophysiol., 122*, 2294–2303.

Finkelstein, A., Fontolan, L., Economo, M. N., Li, N., Romani, S., & Svoboda, K. (2021). Attractor dynamics gate cortical information flow during decision-making. *Nat. Neurosci., 24*, 843–850.

Fonseca, M. S., Murakami, M., & Mainen, Z. F. (2015). Activation of dorsal raphe serotonergic neurons promotes waiting but is not reinforcing. *Curr. Biol., 25*, 306–315.

Forssberg, H., & Grillner, S. (1973). The locomotion of the acute spinal cat injected with clonidine I.V. *Brain Res., 50*, 184–186.

Foster, N. N., Barry, J., Korobkova, L., Garcia, L., Gao, L., Becerra, M., Sherafat, Y., Peng, B., Li, X., Choi, J. H., Gou, L., Zingg, B., Azam, S., Lo, D., Khanjani, N., Zhang, B., Stanis, J., Bowman, I., Cotter, K., Cao, C., Yamashita, S., Tugangui, A., Li, A., Jiang, T, Jia, X., Feng, Z., Aquino, S., Mun, H. S., Zhu, M., Santarelli, A., Benavidez, N. L., Song, M., Dan, G., Fayzullina, M., Ustrell, S., Boesen, T., Johnson, D. L., Xu, H., Bienkowski, M. S., Yang, X. W., Gong, H., Levine, M. S., Wickersham, I., Luo, Q., Hahn, J. D., Lim, B. K., Zhang, L. I., Cepeda, C., Hintiryan, H., & Dong, H. W. (2021). The mouse cortico-basal ganglia-thalamic network. *Nature, 598*, 188–194.

Freud, S. (1878). Uber spinalganglien und ruckenmak des petromyzon. *Sber. Akad. Wiss., 77*, 81–167.

Fried, I., Haggard, P., He, B. J., & Schurger, A. (2017). Volition and action in the human brain: Processes, pathologies, and reasons. *J. Neurosci.*, *37*, 10842–10847.

Fried, I., Katz, A., McCarthy, G., Sass, K. J., Williamson, P., Spencer, S. S., & Spencer, D. D. (1991). Functional organization of human supplementary motor cortex studied by electrical stimulation. *J. Neurosci.*, *11*, 3656–3666.

Friederici, A. D. (2002). Towards a neural basis of auditory sentence processing. *Trends Cogn. Sci.*, *6*, 78–84.

Fritzsch, B., Sonntag, R., Dubuc, R., Ohta, Y., & Grillner, S. (1990). Organization of the six motor nuclei innervating the ocular muscles in lamprey. *J. Comp. Neurol.*, *294*, 491–506.

Frost Nylén, J., Hjorth, J. J. J., Grillner, S., & Hellgren Kotaleski, J. (2021). Dopaminergic and cholinergic modulation of large scale networks in silico using snudda. *Front. Neural Circuits.*, *15*, 748989.

Fuentes, R., Petersson, P., Siesser, W. B., Caron, M. G., & Nicolelis, M. A. (2009). Spinal cord stimulation restores locomotion in animal models of Parkinson's disease. *Science*, *323*, 1578–1582.

Fujita, H., Kodama, T., & du Lac, S. (2020). Modular output circuits of the fastigial nucleus for diverse motor and nonmotor functions of the cerebellar vermis. *Elife*, *9*.

Fukson, O. I., Berkinblit, M. B., & Feldman, A. G. (1980). The spinal frog takes into account the scheme of its body during the wiping reflex. *Science*, *209*, 1261–1263.

Galletti, C., & Fattori, P. (2018). The dorsal visual stream revisited: Stable circuits or dynamic pathways? *Cortex*, *98*, 203–217.

Gariépy, J. F., Missaghi, K., Chevallier, S., Chartre, S., Robert, M., Auclair, F., Lund, J. P., & Dubuc, R. (2012). Specific neural substrate linking respiration to locomotion. *Proc. Natl. Acad. Sci. USA*, *109*, E84–E92.

Garwicz, M., Christensson, M., & Psouni, E. (2009). A unifying model for timing of walking onset in humans and other mammals. *Proc. Natl. Acad. Sci. USA*, *106*, 21889–21893.

Georgopoulos, A. P. (1986). On reaching. *Annu. Rev. Neurosci.*, *9*, 147–170.

Georgopoulos, A. P., DeLong, M. R., & Crutcher, M. D. (1983). Relations between parameters of step-tracking movements and single cell discharge in the globus pallidus and subthalamic nucleus of the behaving monkey. *J. Neurosci.*, *3*, 1586–1598.

Georgopoulos, A. P., & Grillner, S. (1989). Visuomotor coordination in reaching and locomotion. *Science*, *245*, 1209–1210.

Georgopoulos, A. P., Kalaska, J. F., Caminiti, R., & Massey, J. T. (1982). On the relations between the direction of two-dimensional arm movements and cell discharge in primate motor cortex. *J. Neurosci.*, *2*, 1527–1537.

Georgopoulos, A. P., Kettner, R. E., & Schwartz, A. B. (1988). Primate motor cortex and free arm movements to visual targets in three-dimensional space. II. Coding of the direction of movement by a neuronal population. *J. Neurosci.*, *8*, 2928–2937.

Georgopoulos, A. P., Schwartz, A. B., & Kettner, R. E. (1986). Neuronal population coding of movement direction. *Science, 233*, 1416–1419.

Gironell, A., Pascual-Sedano, B., Aracil, I., Marin-Lahoz, J., Pagonabarraga, J., & Kulisevsky, J. (2018). Tremor types in Parkinson disease: A descriptive study using a new classification. *Parkinsons Dis., 2018*, 4327597.

Goetz, C. G. (2011). The history of Parkinson's disease: Early clinical descriptions and neurological therapies. *Cold Spring Harb. Perspect. Med., 1*, a008862.

Goldberg, J. A., & Reynolds, J. N. (2011). Spontaneous firing and evoked pauses in the tonically active cholinergic interneurons of the striatum. *Neuroscience, 198*, 27–43.

González-Rodríguez, P., Zampese, E., Stout, K. A., Guzman, J. N., Ilijic, E., Yang, B., Tkatch, T., Stavarache, M. A., Wokosin, D. L., Gao, L., Kaplitt, M. G., Lopez-Barneo, J., Schumacker, P. T., & Surmeier, D. J. (2021). Disruption of mitochondrial complex I induces progressive Parkinsonism. *Nature, 599*, 650–656.

Goslow, G. E., Jr., Reinking, R. M., & Stuart, D. G. (1973). The cat step cycle: Hind limb joint angles and muscle lengths during unrestrained locomotion. *J. Morphol., 141*, 1–41.

Goulding, M. (2009). Circuits controlling vertebrate locomotion: Moving in a new direction. *Nat. Rev. Neurosci., 10*, 507–518.

Gowers, W. R. (1886). *A manual of diseases of the nervous system II.* J. & A. Churchill.

Granger, A. J., Wallace, M. L., & Sabatini, B. L. (2017). Multi-transmitter neurons in the mammalian central nervous system. *Curr. Opin. Neurobiol., 45*, 85–91.

Gray, J. (1968). *Animal locomotion (the world naturalist).* Littlehampton Book Services Ltd.

Gray, L. A., O'Reilly, J. C., & Nishikawa, K. C. (1997). Evolution of forelimb movement patterns for prey manipulation in anurans. *J. Exp. Zool., 277*, 417–424.

Graybiel, A. M. (1998). The basal ganglia and chunking of action repertoires. *Neurobiol. Learn. Mem., 70*, 119–136.

Graybiel, A. M., & Grafton, S. T. (2015). The striatum: Where skills and habits meet. *Cold Spring Harb. Perspect. Biol., 7*, a021691.

Graybiel, A. M., & Ragsdale, C. W., Jr. (1978). Histochemically distinct compartments in the striatum of human, monkeys, and cat demonstrated by acetylthiocholinesterase staining. *Proc. Natl. Acad. Sci. USA, 75*, 5723–5726.

Graybiel, A. M., & Rauch, S. L. (2000). Toward a neurobiology of obsessive-compulsive disorder. *Neuron, 28*, 343–347.

Graziano, M. S. A. (2016). Ethological action maps: A paradigm shift for the motor cortex. *Trends Cogn. Sci., 20*, 121–132.

Graziano, M. S. A., & Aflalo, T. N. (2007). Mapping behavioral repertoire onto the cortex. *Neuron, 56*, 239–251.

Grillner, S. (1981). Control of locomotion in bipeds, tetrapods, and fish. In V. Brooks (Ed.), *Handbook of physiology* (pp. 1179–1236). American Physiological Society.

Grillner, S. (1985). Neurobiological bases of rhythmic motor acts in vertebrates. *Science, 228*, 143–149.

Grillner, S. (2003). The motor infrastructure: From ion channels to neuronal networks. *Nat. Rev. Neurosci., 4*, 573–586.

Grillner, S. (2012). Fundamentals of motor systems. In L. R. Squire, D. Berg, F. E. Bloom, S. du Lac, A. Ghosh, & N. C. Spitzer (Eds.), *Fundamental neuroscience* (pp. 600–612). Elsevier.

Grillner, S. (2021). Evolution of the vertebrate motor system—From forebrain to spinal cord. *Curr. Opin. Neurobiol., 71*, 11–18.

Grillner, S., Halbertsma, J., Nilsson, J., & Thorstensson, A. (1979). The adaptation to speed in human locomotion. *Brain Res., 165*, 177–182.

Grillner, S., Hellgren, J., Menard, A., Saitoh, K., & Wikstrom, M. A. (2005). Mechanisms for selection of basic motor programs—Roles for the striatum and pallidum. *Trends Neurosci., 28*, 364–370.

Grillner, S., & Kozlov, A. (2021). The CPGs for limbed locomotion-facts and fiction. *Int. J. Mol. Sci., 22*.

Grillner, S., Kozlov, A., Dario, P., Stefanini, C., Menciassi, A., Lansner, A., & Hellgren Kotaleski, J. (2007). Modeling a vertebrate motor system: pattern generation, steering and control of body orientation. *Prog. Brain Res., 165*, 221–234.

Grillner, S., & Lund, S. (1968). The origin of a descending pathway with monosynaptic action on flexor motoneurones. *Acta. Physiol. Scand., 74*, 274–284.

Grillner, S., & El Manira, A. (2020). Current principles of motor control, with special reference to vertebrate locomotion. *Physiol. Rev., 100*, 271–320.

Grillner, S., Markram, H., De Schutter, E., Silberberg, G., & LeBeau, F. E. (2005). Microcircuits in action—from CPGs to neocortex. *Trends Neurosci., 28*, 525–533.

Grillner, S., McClellan, A., & Perret, C. (1981). Entrainment of the spinal pattern generators for swimming by mechano-sensitive elements in the lamprey spinal cord in vitro. *Brain Res., 217*, 380–386.

Grillner, S., Perret, C., & Zangger, P. (1976). Central generation of locomotion in the spinal dogfish. *Brain Res., 109*, 255–269.

Grillner, S., & Robertson, B. (2016). The basal ganglia over 500 million years. *Curr. Biol., 26*, R1088–R1100.

Grillner, S., Robertson, B., & Kotaleski, J. H. (2020). Basal ganglia—a motion perspective. *Compr. Physiol., 10*, 1241–1275.

Grillner, S., Robertson, B., & Stephenson-Jones, M. (2013). The evolutionary origin of the vertebrate basal ganglia and its role in action selection. *J. Physiol., 591*, 5425–5431.

Grillner, S., & Wallén, P. (2004). Innate versus learned movements—a false dichotomy? *Prog. Brain Res.*, *143*, 3–12.

Grillner, S., Williams, T., & Lagerbäck, P. A. (1984). The edge cell, a possible intraspinal mechanoreceptor. *Science*, *223*, 500–503.

Grillner, S., & Zangger, P. (1975). How detailed is the central pattern generation for locomotion? *Brain Res.*, *88*, 367–371.

Guan, N. N., Xu, L., Zhang, T., Huang, C. X., Wang, Z., Dahlberg, E., Wang, H., Wang, F., Pallucchi, I., Hua, Y., El Manira, A., & Song, J. (2021). A specialized spinal circuit for command amplification and directionality during escape behavior. *Proc. Natl. Acad. Sci. USA*, *118*(42), e2106785118.

Ha, N. T., & Dougherty, K. J. (2018). Spinal shox2 interneuron interconnectivity related to function and development. *Elife*, *7*, e42519.

Hägglund, M., Dougherty, K. J., Borgius, L., Itohara, S., Iwasato, T., & Kiehn, O. (2013). Optogenetic dissection reveals multiple rhythmogenic modules underlying locomotion. *Proc. Natl. Acad. Sci. USA*, *110*, 11589–11594.

Hartline, P. H., Kass, L., & Loop, M. S. (1978). Merging of modalities in the optic tectum: Infrared and visual integration in rattlesnakes. *Science*, *199*, 1225–1229.

Haynes, W. I., & Haber, S. N. (2013). The organization of prefrontal-subthalamic inputs in primates provides an anatomical substrate for both functional specificity and integration: Implications for basal ganglia models and deep brain stimulation. *J. Neurosci.*, *33*, 4804–4814.

Henningsen, J. B., Soylu-Kucharz, R., Bjorkqvist, M., & Petersen, A. (2021). Effects of excitotoxicity in the hypothalamus in transgenic mouse models of Huntington disease. *Heliyon*, *7*, e07808.

Herculano-Houzel, S., Collins, C. E., Wong, P., Kaas, J. H., & Lent, R. (2008). The basic nonuniformity of the cerebral cortex. *Proc. Natl. Acad. Sci. USA*, *105*, 12593–12598.

Higurashi, Y., Maier, M. A., Nakajima, K., Morita, K., Fujiki, S., Aoi, S., Mori, F., Murata, A., & Inase, M. (2019). Locomotor kinematics and EMG activity during quadrupedal versus bipedal gait in the Japanese macaque. *J. Neurophysiol.*, *122*, 398–412.

Hikosaka, O., & Wurtz, R. H. (1983). Visual and oculomotor functions of monkey substantia nigra pars reticulata. IV. Relation of substantia nigra to superior colliculus. *J. Neurophysiol.*, *49*, 1285–1301.

Hjorth, J. J. J., Blackwell, K. T., & Kotaleski, J. H. (2009). Gap junctions between striatal fast-spiking interneurons regulate spiking activity and synchronization as a function of cortical activity. *J. Neurosci.*, *29*, 5276–5286.

Hjorth, J. J. J., Kozlov, A., Carannante, I., Frost Nylén, J., Lindroos, R., Johansson, Y., Tokarska, A., Dorst, M. C., Suryanarayana, S. M., Silberberg, G., Hellgren Kotaleski, J., & Grillner, S. (2020). The microcircuits of striatum in silico. *Proc. Natl. Acad. Sci. USA*, *117*, 9554–9565.

Holly, E. N., Davatolhagh, M. F., Choi, K., Alabi, O. O., Vargas Cifuentes, L., & Fuccillo, M. V. (2019). Striatal low-threshold spiking interneurons regulate goal-directed learning. *Neuron*, *103*, 92–101 e106.

Hou, X. H., Hyun, M., Taranda, J., Huang, K. W., Todd, E., Feng, D., Atwater, E., Croney, D., Zeidel, M. L., Osten, P., & Sabatini, B. L. (2016). Central control circuit for context-dependent micturition. *Cell*, *167*, 73–86 e12.

Howe, M., Ridouh, I., Allegra Mascaro, A. L., Larios, A., Azcorra, M., & Dombeck, D. A. (2019). Coordination of rapid cholinergic and dopaminergic signaling in striatum during spontaneous movement. *Elife*, *8*, e44903.

Hu, H., Cui, Y., & Yang, Y. (2020). Circuits and functions of the lateral habenula in health and in disease. *Nat. Rev. Neurosci.*, *21*, 277–295.

Hubel, D. H., & Wiesel, T. N. (1969). Anatomical demonstration of columns in the monkey striate cortex. *Nature*, *221*, 747–750.

Hunger, L., Kumar, A., & Schmidt, R. (2020). Abundance compensates kinetics: Similar effect of dopamine signals on D1 and D2 receptor populations. *J. Neurosci.*, *40*, 2868–2881.

Iigaya, K., Fonseca, M. S., Murakami, M., Mainen, Z. F., & Dayan, P. (2018). An effect of serotonergic stimulation on learning rates for rewards apparent after long intertrial intervals. *Nat. Commun.*, *9*(1), *2477*.

Isa, T., Marquez-Legorreta, E., Grillner, S., & Scott, E. K. (2021). The tectum/superior colliculus as the vertebrate solution for spatial sensory integration and action. *Curr. Biol.*, *31*, R741–R762.

Ito, M. (1972). Neural design of the cerebellar motor control system. *Brain Res.*, *40*, 81–84.

Ito, M. (1984). *The cerebellum and neural control*. Raven Press Books Ltd.

Ito, M. (1989). Long-term depression. *Annu. Rev. Neurosci.*, *12*, 85–102.

Ito, M. (2002). The molecular organization of cerebellar long-term depression. *Nat. Rev. Neurosci.*, *3*, 896–902.

Ito, M., & Yoshida, M. (1966). The origin of cerebral-induced inhibition of Deiters neurones. I. Monosynaptic initiation of the inhibitory postsynaptic potentials. *Exp. Brain Res.*, *2*, 330–349.

Iwaniuk, A. N., & Whishaw, I. Q. (2000). On the origin of skilled forelimb movements. *Trends Neurosci.*, *23*, 372–376.

Jalalvand, E., Alvelid, J., Coceano, G., Edwards, S., Robertson, B., Grillner, S., & Testa, I. (2022). ExSTED microscopy reveals contrasting functions of dopamine and somatostatin CSF-c neurons along the lamprey central canal. *Elife*, *11*, e73114.

Jalalvand, E., Robertson, B., Tostivint, H., Wallen, P., & Grillner, S. (2016). The spinal cord has an intrinsic system for the control of pH. *Curr. Biol.*, *26*, 1346–1351.

Janak, P. H., & Tye, K. M. (2015). From circuits to behaviour in the amygdala. *Nature*, *517*, 284–292.

Jay, M. F., & Sparks, D. L. (1984). Auditory receptive fields in primate superior colliculus shift with changes in eye position. *Nature*, *309*, 345–347.

Jean, A. (1990). Brainstem control of swallowing: Localization and organization of the central pattern generator for swallowing. In A. Taylor (Ed.), *Neurophysiology of the jaws and teeth* (pp. 294–321). MacMillan.

Jean, A. (2001). Brain stem control of swallowing: Neuronal network and cellular mechanisms. *Physiol. Rev., 81*, 929–969.

Jessell, T. M. (2000). Neuronal specification in the spinal cord: Inductive signals and transcriptional codes. *Nat. Rev. Genet., 1*, 20–29.

Jin, X., Tecuapetla, F., & Costa, R. M. (2014). Basal ganglia subcircuits distinctively encode the parsing and concatenation of action sequences. *Nat. Neurosci., 17*, 423–430.

Jirenhed, D. A., Rasmussen, A., Johansson, F., & Hesslow, G. (2017). Learned response sequences in cerebellar Purkinje cells. *Proc. Natl. Acad. Sci. USA, 114*, 6127–6132.

Johansson, R. S., & Flanagan, J. R. (2009). Coding and use of tactile signals from the fingertips in object manipulation tasks. *Nat. Rev. Neurosci., 10*, 345–359.

Johansson, Y., & Silberberg, G. (2020). The functional organization of cortical and thalamic inputs onto five types of striatal neurons is determined by source and target cell identities. *Cell Rep., 30*, 1178–1194 e1173.

Johnels, B., Ingvarsson, P. E., Steg, G., & Olsson, T. (2001). The Posturo-Locomotion-Manual Test. A simple method for the characterization of neurological movement disturbances. *Adv. Neurol., 87*, 91–100.

Jones, E. G. (2000). Microcolumns in the cerebral cortex. *Proc. Natl. Acad. Sci. USA, 97*, 5019–5021.

Jones, M. R., Grillner, S., & Robertson, B. (2009). Selective projection patterns from subtypes of retinal ganglion cells to tectum and pretectum: Distribution and relation to behavior. *J. Comp. Neurol., 517*, 257–275.

Jörntell, H., & Ekerot, C. F. (2011). Receptive field remodeling induced by skin stimulation in cerebellar neurons in vivo. *Front. Neural Circuits, 5*, 3.

Jung, H., Baek, M., D'Elia, K. P., Boisvert, C., Currie, P. D., Tay, B. H., Venkatesh, B., Brown, S. M., Heguy, A., Schoppik, D., & Dasen, J. S. (2018). The ancient origins of neural substrates for land walking. *Cell, 172*, 667–682 e615.

Kaas, J. H., Qi, H. X., & Stepniewska, I. (2022). Escaping the nocturnal bottleneck, and the evolution of the dorsal and ventral streams of visual processing in primates. *Philos. Trans. R. Soc. Lond. B. Biol. Sci., 377*(1844), 20210293.

Kably, B., & Drew, T. (1998a). Corticoreticular pathways in the cat. I. Projection patterns and collaterization. *J. Neurophysiol., 80*, 389–405.

Kably, B., & Drew, T. (1998b). Corticoreticular pathways in the cat. II. Discharge activity of neurons in area 4 during voluntary gait modifications. *J. Neurophysiol., 80*, 406–424.

Kalueff, A. V., Stewart, A. M., Song, C., Berridge, K. C., Graybiel, A. M., & Fentress, J. C. (2016). Neurobiology of rodent self-grooming and its value for translational neuroscience. *Nat. Rev. Neurosci.*, *17*, 45–59.

Kardamakis, A. A., Pérez-Fernández, J., & Grillner, S. (2016). Spatiotemporal interplay between multisensory excitation and recruited inhibition in the lamprey optic tectum. *Elife*, *5*, e16472.

Kardamakis, A. A., Saitoh, K., & Grillner, S. (2015). Tectal microcircuit generating visual selection commands on gaze-controlling neurons. *Proc. Natl. Acad. Sci. USA*, *112*, E1956–1965.

Karigo, T., Kennedy, A., Yang, B., Liu, M., Tai, D., Wahle, I. A., & Anderson, D. J. (2021). Distinct hypothalamic control of same- and opposite-sex mounting behaviour in mice. *Nature*, *589*, 258–263.

Karube, F., Takahashi, S., Kobayashi, K., & Fujiyama, F. (2019). Motor cortex can directly drive the globus pallidus neurons in a projection neuron type-dependent manner in the rat. *Elife*, *8*, e49511.

Kashin, S. M., Malinin, L. K., Orlovsky, G. N., & Poddubny, A. G. (1977). Behavior of some fishes during hunting (in Russian). *Zoological Journal*, *56*, 1328–1338.

Kato, R., Takaura, K., Ikeda, T., Yoshida, M., & Isa, T. (2011). Contribution of the retino-tectal pathway to visually guided saccades after lesion of the primary visual cortex in monkeys. *Eur. J. Neurosci.*, *33*, 1952–1960.

Kawai, R., Markman, T., Poddar, R., Ko, R., Fantana, A. L., Dhawale, A. K., Kampff, A. R., & Olveczky, B. P. (2015). Motor cortex is required for learning but not for executing a motor skill. *Neuron*, *86*, 800–812.

Kawato, M., Furukawa, K., & Suzuki, R. (1987). A hierarchical neural-network model for control and learning of voluntary movement. *Biol. Cybern.*, *57*, 169–185.

Kawato, M., & Gomi, H. (1992). The cerebellum and VOR/OKR learning models. *Trends Neurosci.*, *15*, 445–453.

Kawato, M., Ohmae, S., Hoang, H., & Sanger, T. (2021). 50 years since the Marr, Ito, and Albus models of the cerebellum. *Neuroscience*, *462*, 151–174.

Ketzef, M., & Silberberg, G. (2021). Differential synaptic input to external globus pallidus neuronal subpopulations in vivo. *Neuron*, *109*, 516–529 e514.

Kiehn, O. (2016). Decoding the organization of spinal circuits that control locomotion. *Nat. Rev. Neurosci.*, *17*, 224–238.

Kim, H. F., & Hikosaka, O. (2015). Parallel basal ganglia circuits for voluntary and automatic behaviour to reach rewards. *Brain*, *138*, 1776–1800.

Klaus, A., Planert, H., Hjorth, J. J., Berke, J. D., Silberberg, G., & Kotaleski, J. H. (2011). Striatal fast-spiking interneurons: from firing patterns to postsynaptic impact. *Front. Syst. Neurosci.*, *5*, 57.

Kleinfeld, D., Deschenes, M., Wang, F., & Moore, J. D. (2014). More than a rhythm of life: Breathing as a binder of orofacial sensation. *Nat. Neurosci.*, *17*, 647–651.

Knudsen, E. I. (1982). Auditory and visual maps of space in the optic tectum of the owl. *J. Neurosci.*, *2*, 1177–1194.

Knudsen, E. I., & Konishi, M. (1978). A neural map of auditory space in the owl. *Science*, *200*, 795–797.

Kohl, J., Babayan, B. M., Rubinstein, N. D., Autry, A. E., Marin-Rodriguez, B., Kapoor, V., Miyamishi, K., Zweifel, L. S., Luo, L., Uchida, N., & Dulac, C. (2018). Functional circuit architecture underlying parental behaviour. *Nature*, *556*, 326–331.

Kolta, A., Brocard, F., Verdier, D., & Lund, J. P. (2007). A review of burst generation by trigeminal main sensory neurons. *Arch. Oral Biol.*, *52*, 325–328.

Kotaleski, J. H., Lansner, A., & Grillner, S. (1999). Neural mechanisms potentially contributing to the intersegmental phase lag in lamprey. II. Hemisegmental oscillations produced by mutually coupled excitatory neurons. *Biol. Cybern.*, *81*, 299–315.

Kozlov, A., Huss, M., Lansner, A., Kotaleski, J. H., & Grillner, S. (2009). Simple cellular and network control principles govern complex patterns of motor behavior. *Proc. Natl. Acad. Sci. USA*, *106*, 20027–20032.

Kozlov, A. K., Kardamakis, A. A., Hellgren Kotaleski, J., & Grillner, S. (2014). Gating of steering signals through phasic modulation of reticulospinal neurons during locomotion. *Proc. Natl. Acad. Sci. USA*, *111*, 3591–3596.

Kumar, S., & Hedges, S. B. (1998). A molecular timescale for vertebrate evolution. *Nature*, *392*, 917–920.

Kunwar, P. S., Zelikowsky, M., Remedios, R., Cai, H., Yilmaz, M., Meister, M., & Anderson, D. J. (2015). Ventromedial hypothalamic neurons control a defensive emotion state. *Elife*, *4*.

Lacalli, T. (2022). An evolutionary perspective on chordate brain organization and function: Insights from amphioxus, and the problem of sentience. *Philos. Trans. R. Soc. Lond. B. Biol. Sci.*, *377*(1844), 20200520.

Lacey, C. J., Boyes, J., Gerlach, O., Chen, L., Magill, P. J., & Bolam, J. P. (2005). GABA(B) receptors at glutamatergic synapses in the rat striatum. *Neuroscience*, *136*, 1083–1095.

Lamanna, F., Hervas-Sotomayor, F., Oel, A., Jandzik, D., Sobrido-Cameán, D., Martik, M. L., Green, S. A., Brüning, T., Mössinger, K., Schmidt, J., Schneider, C., Sepp, M., Murat, F., Smith, J. J., Bronner, M. E., Rodicio, C., Barreiro-Iglesias, A., Medeiros, D. M., Arendt, D., & Kaessmann, H. (2022). Reconstructing the ancestral vertebrate brain using a lamprey neural cell type atlas. *bioRXiv*. https://doi.org/10.1101/2022.02.28.482278

Lang, I. M., Sarna, S. K., & Dodds, W. J. (1993). Pharyngeal, esophageal, and proximal gastric responses associated with vomiting. *Am. J. Physiol.*, *265*, G963–G972.

Lawrence, D. G., & Kuypers, H. G. (1968a). The functional organization of the motor system in the monkey. I. The effects of bilateral pyramidal lesions. *Brain*, *91*, 1–14.

Lawrence, D. G., & Kuypers, H. G. (1968b). The functional organization of the motor system in the monkey. II. The effects of lesions of the descending brain-stem pathways. *Brain*, *91*, 15–36.

Lazaridis, I., Tzortzi, O., Weglage, M., Martin, A., Xuan, Y., Parent, M., Johansson, Y., Fuzik, J., Furth, D., Fenno, L. E., Ramakrishnan, C., Silberberg, G., Deisseroth, K., Carlen, M., & Meletis, K. (2019). A hypothalamus-habenula circuit controls aversion. *Mol. Psychiatry*, *24*, 1351–1368.

Lecca, S., Meye, F. J., Trusel, M., Tchenio, A., Harris, J., Schwarz, M. K., Burdakov, D., Georges, F., & Mameli, M. (2017). Aversive stimuli drive hypothalamus-to-habenula excitation to promote escape behavior. *Elife*, *6*, e30697.

LeDoux, J. (2012). Rethinking the emotional brain. *Neuron*, *73*, 653–676.

Lefler, Y., Campagner, D., & Branco, T. (2020). The role of the periaqueductal gray in escape behavior. *Curr. Opin. Neurobiol.*, *60*, 115–121.

Lemon, R. N. (2008). Descending pathways in motor control. *Annu. Rev. Neurosci.*, *31*, 195–218.

Lemon, S. S. F., & Sherrington, C. S. (1917). Observations on the excitable cortex of the chimpanzee, orangutan and gorilla. *Q. J. Exp. Physiol.*, *11,* 135–222.

Li, P., Janczewski, W. A., Yackle, K., Kam, K., Pagliardini, S., Krasnow, M. A., & Feldman, J. L. (2016). The peptidergic control circuit for sighing. *Nature*, *530*, 293–297.

Liddell, E. G. T., & Phillips, C. G. (1944). Pyramidal section in the cat. *Brain*, *67*, 1–9.

Lilascharoen, V., Wang, E. H., Do, N., Pate, S. C., Tran, A. N., Yoon, C. D., Choi, J. H., Wang, X. Y., Pribiag, H., Park, Y. G., Chung, K., & Lim, B. K. (2021). Divergent pallidal pathways underlying distinct Parkinsonian behavioral deficits. *Nat. Neurosci.*, *24*, 504–515.

Lin, D., Boyle, M. P., Dollar, P., Lee, H., Lein, E. S., Perona, P., & Anderson, D. J. (2011). Functional identification of an aggression locus in the mouse hypothalamus. *Nature*, *470*, 221–226.

Lindahl, M., Kamali Sarvestani, I., Ekeberg, O., & Kotaleski, J. H. (2013). Signal enhancement in the output stage of the Basal Ganglia by synaptic short-term plasticity in the direct, indirect, and hyperdirect pathways. *Front. Comput. Neurosci.*, *7*, 76. doi: 10.3389/fncom .2013.00076.

Lindroos, R., Dorst, M. C., Du, K., Filipovic, M., Keller, D., Ketzef, M., Kozlov, A. K., Kumar, A., Lindahl, M., Nair, A. G., Pérez-Fernández, J., Grillner, S., Silberberg, G., & Hellgren Kotaleski, J. (2018). Basal Ganglia neuromodulation over multiple temporal and structural scales-simulations of direct pathway MSNs investigate the fast onset of dopaminergic effects and predict the role of Kv4.2. *Front. Neural Circuits*, *12*, 3, doi: 10.3389/fncir.2018.00003.

Lisberger, S. G. (2021). The rules of cerebellar learning: Around the Ito hypothesis. *Neuroscience*, *462*, 175–190.

Llinás, R. R. (2013). The olivo-cerebellar system: A key to understanding the functional significance of intrinsic oscillatory brain properties. *Front. Neural Circuits*, *7*, 96.

Lo, L., Yao, S., Kim, D. W., Cetin, A., Harris, J., Zeng, H., Anderson, D. J., & Weissbourd, B. (2019). Connectional architecture of a mouse hypothalamic circuit node controlling social behavior. *Proc. Natl. Acad. Sci. USA*, *116*, 7503–7512.

Maier, M., Ballester, B. R., Duff, A., Duarte Oller, E., & Verschure, P. F. M. J. (2019). Effect of specific over nonspecific VR-based rehabilitation on poststroke motor recovery: A systematic meta-analysis. *Neurorehabil. Neural Repair, 33*, 112–129.

Mallet, N., Micklem, B. R., Henny, P., Brown, M. T., Williams, C., Bolam, J. P., Nakamura, K. C., & Magill, P. J. (2012). Dichotomous organization of the external globus pallidus. *Neuron, 74*, 1075–1086.

Mallet, N., Schmidt, R., Leventhal, D., Chen, F., Amer, N., Boraud, T., & Berke, J. D. (2016). Arkypallidal cells send a stop signal to striatum. *Neuron, 89*, 308–316.

Mandelbaum, G., Taranda, J., Haynes, T. M., Hochbaum, D. R., Huang, K. W., Hyun, M., Umadevi Venkataraju, K., Straub, C., Wang, W., Robertson, K., Osten, P., & Sabatini, B. L. (2019). Distinct cortical-thalamic-striatal circuits through the parafascicular nucleus. *Neuron, 102*, 636–652 e637.

Marigold, D. S., & Drew, T. (2017). Posterior parietal cortex estimates the relationship between object and body location during locomotion. *Elife, 6*.

Markram, H., Toledo-Rodriguez, M., Wang, Y., Gupta, A., Silberberg, G., & Wu, C. (2004). Interneurons of the neocortical inhibitory system. *Nat. Rev. Neurosci., 5*, 793–807.

Marr, D. (1969). A theory of cerebellar cortex. *J. Physiol., 202*, 437–470.

Märtin, A., Calvigioni, D., Tzortzi, O., Fuzik, J., Wärnberg, E., & Meletis, K. (2019). A spatio-molecular map of the striatum. *Cell Rep., 29*, 4320–4333 e4325.

Martinez-Gonzalez, C., Bolam, J. P., & Mena-Segovia, J. (2011). Topographical organization of the pedunculopontine nucleus. *Front. Neuroanat., 5*, 22.

Masullo, L., Mariotti, L., Alexandre, N., Freire-Pritchett, P., Boulanger, J., & Tripodi, M. (2019). Genetically defined functional modules for spatial orienting in the mouse superior colliculus. *Curr. Biol., 29*, 2892–2904 e2898.

Matsuyama, K., Mori, F., Nakajima, K., Drew, T., Aoki, M., & Mori, S. (2004). Locomotor role of the corticoreticular-reticulospinal-spinal interneuronal system. *Prog. Brain Res., 143*, 239–249.

McCormick, D. A., & Thompson, R. F. (1984a). Cerebellum: Essential involvement in the classically conditioned eyelid response. *Science, 223*, 296–299.

McCormick, D. A., & Thompson, R. F. (1984b). Neuronal responses of the rabbit cerebellum during acquisition and performance of a classically conditioned nictitating membrane-eyelid response. *J. Neurosci., 4*, 2811–2822.

McElvain, L. E., Chen, Y., Moore, J. D., Brigidi, G. S., Bloodgood, B. L., Lim, B. K., Costa, R. M., & Kleinfeld, D. (2021). Specific populations of basal ganglia output neurons target distinct brain stem areas while collateralizing throughout the diencephalon. *Neuron, 109*, 1721–1738 e1724.

Meredith, M. A., & Stein, B. E. (1986). Spatial factors determine the activity of multisensory neurons in cat superior colliculus. *Brain Res., 365*, 350–354.

Mesulam, M. M., Mufson, E. J., Levey, A. I., & Wainer, B. H. (1984). Atlas of cholinergic neurons in the forebrain and upper brainstem of the macaque based on monoclonal choline acetyltransferase immunohistochemistry and acetylcholinesterase histochemistry. *Neuroscience, 12*, 669–686.

Missaghi, K., Le Gal, J. P., Gray, P. A., & Dubuc, R. (2016). The neural control of respiration in lampreys. *Respir. Physiol. Neurobiol., 234*, 14–25.

Miyazaki, K. W., Miyazaki, K., Tanaka, K. F., Yamanaka, A., Takahashi, A., Tabuchi, S., & Doya, K. (2014). Optogenetic activation of dorsal raphe serotonin neurons enhances patience for future rewards. *Curr. Biol., 24*, 2033–2040.

Mizuno, Y., Hattori, N., Mori, H., Suzuki, T., & Tanaka, K. (2001). Parkin and Parkinson's disease. *Curr. Opin. Neurol., 14*, 477–482.

Mohan, H., An, X., Musall, S., Mitra, P. P., Churchland, A. K., & Huang, J. Z. (2019). Pyramidal neuron types and cortical networks, encoding sensorimotor activity coordinating hand-mouth synergy. *Society for Neuroscience Abstract, 227*, 15.

Moore, J. D., Deschenes, M., Furuta, T., Huber, D., Smear, M. C., Demers, M., & Kleinfeld, D. (2013). Hierarchy of orofacial rhythms revealed through whisking and breathing. *Nature, 497*, 205–210.

Moore, J. D., Kleinfeld, D., & Wang, F. (2014). How the brainstem controls orofacial behaviors comprised of rhythmic actions. *Trends Neurosci., 37*, 370–380.

Morgenstern, N. A., Isidro, A. F., Israely, I., & Costa, R. M. (2022). Pyramidal tract neurons drive amplification of excitatory inputs to striatum through cholinergic interneurons. *Sci. Adv., 8*(6), eabh4315.

Morquette, P., Verdier, D., Kadala, A., Fethiere, J., Philippe, A. G., Robitaille, R., & Kolta, A. (2015). An astrocyte-dependent mechanism for neuronal rhythmogenesis. *Nat. Neurosci., 18*, 844–854.

Mountcastle, V. B., Davies, P. W., & Berman, A. L. (1957). Response properties of neurons of cat's somatic sensory cortex to peripheral stimuli. *J. Neurophysiol., 20*, 374–407.

Nachev, P., Kennard, C., & Husain, M. (2008). Functional role of the supplementary and pre-supplementary motor areas. *Nat. Rev. Neurosci., 9*, 856–869.

Nair, A. G., Castro, L. R. V., El Khoury, M., Gorgievski, V., Giros, B., Tzavara, E. T., Hellgren-Kotaleski, J., & Vincent, P. (2019). The high efficacy of muscarinic M4 receptor in D1 medium spiny neurons reverses striatal hyperdopaminergia. *Neuropharmacology, 146*, 74–83.

Nair, A. G., Gutierrez-Arenas, O., Eriksson, O., Vincent, P., & Hellgren Kotaleski, J. (2015). Sensing positive versus negative reward signals through adenylyl cyclase-coupled GPCRs in direct and indirect pathway striatal medium spiny neurons. *J. Neurosci., 35*, 14017–14030.

Nakajima, T., Fortier-Lebel, N., & Drew, T. (2019). Premotor cortex provides a substrate for the temporal transformation of information during the planning of gait modifications. *Cereb. Cortex., 29*, 4982–5008.

Nambu, A., Takada, M., Inase, M., & Tokuno, H. (1996). Dual somatotopical representations in the primate subthalamic nucleus: Evidence for ordered but reversed body-map transformations from the primary motor cortex and the supplementary motor area. *J. Neurosci., 16*, 2671–2683.

Nambu, A., Tokuno, H., & Takada, M. (2002). Functional significance of the cortico-subthalamo-pallidal "hyperdirect" pathway. *Neurosci. Res., 43*, 111–117.

Nelson, A., Abdelmesih, B., & Costa, R. M. (2021). Corticospinal populations broadcast complex motor signals to coordinated spinal and striatal circuits. *Nat. Neurosci., 24*, 1721–1732.

Newman, E. A., & Hartline, P. H. (1981). Integration of visual and infrared information in bimodal neurons in the rattlesnake optic tectum. *Science, 213*, 789–791.

Nieuwenhuys, R., & Nicholson, C. (1998). Lampreys, petromyzontoidea. In R. Nieuwenhuys, H. J. T. Donkelaar, and C. Nicholson (Eds.), *The central nervous system of vertebrates* (pp. 397–495). Springer.

Nonomura, S., Nishizawa, K., Sakai, Y., Kawaguchi, Y., Kato, S., Uchigashima, M., Watanabe, M., Yamanaka, K., Enomoto, K., Chiken, S., Sano, H., Soma, S., Yoshida, J., Samejima, K., Ogawa, M., Kobayashi, K., Nambu, A., Isomura, Y., & Kimura, M. (2018). Monitoring and updating of action selection for goal-directed behavior through the striatal direct and indirect pathways. *Neuron, 99*, 1302–1314 e1305.

Ocana, F. M., Suryanarayana, S. M., Saitoh, K., Kardamakis, A. A., Capantini, L., Robertson, B., & Grillner, S. (2015). The lamprey pallium provides a blueprint of the mammalian motor projections from cortex. *Curr. Biol., 25*, 413–423.

Ohta, Y., & Grillner, S. (1989). Monosynaptic excitatory amino acid transmission from the posterior rhombencephalic reticular nucleus to spinal neurons involved in the control of locomotion in lamprey. *J. Neurophysiol., 62*, 1079–1089.

Olson, I., Suryanarayana, S. M., Robertson, B., & Grillner, S. (2017). Griseum centrale, a homologue of the periaqueductal gray in the lamprey. *IBRO Rep., 2*, 24–30.

Oorschot, D. E. (1996). Total number of neurons in the neostriatal, pallidal, subthalamic, and substantia nigral nuclei of the rat basal ganglia: A stereological study using the cavalieri and optical disector methods. *J. Comp. Neurol., 366*, 580–599.

Orlovsky, G. N. (1969). Spontaneous and induced locomotion of the thalamic cat. *Biofizika, 14*, 1154–1162.

Orlovsky, G. N. (1972a). Activity of rubrospinal neurons during locomotion. *Brain Res., 46*, 99–112.

Orlovsky, G. N. (1972b). Activity of vestibulospinal neurons during locomotion. *Brain Res., 46*, 85–98.

Orlovsky, G. N. (1972c). The effect of different descending systems on flexor and extensor activity during locomotion. *Brain Res., 40*, 359–371.

Orlovsky, G. N., Deliagina, T. G., & Grillner, S. (1999). *Neuronal control of locomotion. From mollusc to man.* Oxford University Press.

Oscarsson, O. (1968). Termination and functional organization of the ventral spino-olivocerebellar path. *J. Physiol., 196*, 453–478.

Oscarsson, O. (1976). Spatial distribution of climbing and mossy fibre inputs into the cerebellar cortex. In O. Creutzfeldt (Ed.), *Afferent and intrinsic organization of laminated structures in the brain* (pp. 34–42). Springer-Verlag.

Papathanou, M., Creed, M., Dorst, M. C., Bimpsidis, Z., Dumas, S., Pettersson, H., Bellone, C., Silberberg, G., Lüscher, C., & Wallén-Mackenzie, A. (2018). Targeting VGLUT2 in mature dopamine neurons decreases mesoaccumbal glutamatergic transmission and identifies a role for glutamate co-release in synaptic plasticity by increasing baseline AMPA/NMDA ratio. *Front. Neural Circuits*, *12*, 64.

Parievsky, A., Moore, C., Kamdjou, T., Cepeda, C., Meshul, C. K., & Levine, M. S. (2017). Differential electrophysiological and morphological alterations of thalamostriatal and corticostriatal projections in the R6/2 mouse model of Huntington's disease. *Neurobiol. Dis.*, *108*, 29–44.

Park, J., Phillips, J. W., Guo, J. Z., Martin, K. A., Hantman, A. W., & Dudman, J. T. (2022). Motor cortical output for skilled forelimb movement is selectively distributed across projection neuron classes. *Sci. Adv.* 8(10), eabj5167.

Parker, J. G., Marshall, J. D., Ahanonu, B., Wu, Y. W., Kim, T. H., Grewe, B. F., Zhang, Y., Li, J. Z., Ding, J. B., Ehlers, M. D., & Schnitzer, M. J. (2018). Diametric neural ensemble dynamics in parkinsonian and dyskinetic states. *Nature. 557*(7704):177–182.

Penfield, W., & Rasmussen, T. (1950). *The cerebral cortex of man: A clinical study of localization of function*. Macmillan.

Peng, H., Xie, P., Liu, L., Kuang, X., Wang, Y., Qu, L., Gong, H., Jiang, S., Li, A., Ruan, Z., Ding, L., Yao, Z., Chen, C., Chen, M., Daigle, T. L., Dalley, R., Ding, Z., Duan, Y., Feiner, A., He, P., Hill, C., Hirokawa, K. E., Hong, G., Huang, L., Kebede, S., Kuo, H. C., Larsen, R., Lesnar, P., Li, L., Li, Q., Li, X., Li, Y., Li, Y., Liu, A., Lu, D., Mok, S., Ng, L., Nguyen, T. N., Ouyang, Q., Pan, J., Shen, E., Song, Y., Sunkin, S. M., Tasic, B., Veldman, M. B., Wakeman, W., Wan, W., Wang, P., Wang, Q., Wang, T., Wang, Y., Xiong, F., Xiong, W., Xu, W., Ye, M., Yin, L., Yu, Y., Yuan, J., Yuan, J., Yun, Z., Zeng, S., Zhang, S., Zhao, S., Zhao, Z., Zhou, Z., Huang, Z. J., Esposito, L., Hawrylycz, M. J., Sorensen, S. A., Yang, X. W., Zheng, Y., Gu, Z., Xie, W., Koch, C., Luo, Q., Harris, J. A., Wang, Y., & Zeng, H. (2021). Morphological diversity of single neurons in molecularly defined cell types. *Nature*, *598*, 174–181.

Pérez-Fernández, J., Kardamakis, A. A., Suzuki, D. G., Robertson, B., & Grillner, S. (2017). Direct dopaminergic projections from the SNc modulate visuomotor transformation in the lamprey tectum. *Neuron*, *96*, 910–924 e915.

Pérez-Fernández, J., Stephenson-Jones, M., Suryanarayana, S. M., Robertson, B., & Grillner, S. (2014). Evolutionarily conserved organization of the dopaminergic system in lamprey: SNc/VTA afferent and efferent connectivity and D2 receptor expression. *J. Comp. Neurol.*, *522*, 3775–3794.

Peron, S., Pancholi, R., Voelcker, B., Wittenbach, J. D., Olafsdottir, H. F., Freeman, J., & Svoboda, K. (2020). Recurrent interactions in local cortical circuits. *Nature*, *579*, 256–259.

Pfaff, D. W. (2017). *How the vertebrate brain regulates behavior*. Harvard University Press.

Philipp, R., & Hoffmann, K. P. (2014). Arm movements induced by electrical microstimulation in the superior colliculus of the macaque monkey. *J. Neurosci.*, *34*, 3350–3363.

Phillips, C. G., & Porter, R. (1977). Corticospinal neurones. Their role in movement. *Monogr. Physiol. Soc.*, *34*, v–xii, 1–450.

Picton, L. D., Bertuzzi, M., Pallucchi, I., Fontanel, P., Dahlberg, E., Bjornfors, E. R., Iacoviello, F., Shearing, P. R., & El Manira, A. (2021). A spinal organ of proprioception for integrated motor action feedback. *Neuron*, *109*, 1188–1201 e1187.

Pivetta, C., Esposito, M. S., Sigrist, M., & Arber, S. (2014). Motor-circuit communication matrix from spinal cord to brainstem neurons revealed by developmental origin. *Cell*, *156*, 537–548.

Planert, H., Szydlowski, S. N., Hjorth, J. J., Grillner, S., & Silberberg, G. (2010). Dynamics of synaptic transmission between fast-spiking interneurons and striatal projection neurons of the direct and indirect pathways. *J. Neurosci.*, *30*, 3499–3507.

Porter, R. (1987). The Florey lecture, 1987. Corticomotoneuronal projections: Synaptic events related to skilled movement. *Proc. R. Soc. Lond. B. Biol. Sci.*, *231*, 147–168.

Puelles, L., & Rubenstein, J. L. (2015). A new scenario of hypothalamic organization: Rationale of new hypotheses introduced in the updated prosomeric model. *Front. Neuroanat.*, *9*, 27.

Rakic, P. (2008). Confusing cortical columns. *Proc. Natl. Acad. Sci. USA*, *105*, 12099–12100.

Rancic, V., & Gosgnach, S. (2021). Recent insights into the rhythmogenic core of the locomotor CPG. *Int. J. Mol. Sci.*, *22*.

Raz, A., Feingold, A., Zelanskaya, V., Vaadia, E., & Bergman, H. (1996). Neuronal synchronization of tonically active neurons in the striatum of normal and parkinsonian primates. *J. Neurophysiol.*, *76*, 2083–2088.

Redgrave, P., & Gurney, K. (2006). The short-latency dopamine signal: a role in discovering novel actions? *Nat. Rev. Neurosci.*, *7*, 967–975.

Redgrave, P., Rodriguez, M., Smith, Y., Rodriguez-Oroz, M. C., Lehericy, S., Bergman, H., Agid, Y., DeLong, M. R., & Obeso, J. A. (2010). Goal-directed and habitual control in the basal ganglia: Implications for Parkinson's disease. *Nat. Rev. Neurosci.*, *11*, 760–772.

Reig, R., & Silberberg, G. (2014). Multisensory integration in the mouse striatum. *Neuron*, *83*, 1200–1212.

Reynolds, J. N. J., Avvisati, R., Dodson, P. D., Fisher, S. D., Oswald, M. J., Wickens, J. R., & Zhang, Y. F. (2022). Coincidence of cholinergic pauses, dopaminergic activation and depolarisation of spiny projection neurons drives synaptic plasticity in the striatum. *Nat. Commun.*, *13*(1), *1296*.

Rice, M. E., & Cragg, S. J. (2004). Nicotine amplifies reward-related dopamine signals in striatum. *Nat. Neurosci.*, *7*, 583–584.

Rietdijk, C. D., Perez-Pardo, P., Garssen, J., van Wezel, R. J., & Kraneveld, A. D. (2017). Exploring Braak's hypothesis of Parkinson's disease. *Front. Neurol.*, *8*, 37.

Rizzolatti, G., & Sinigaglia, C. (2010). The functional role of the parieto-frontal mirror circuit: Interpretations and misinterpretations. *Nat. Rev. Neurosci.*, *11*, 264–274.

Robertson, B., Sengul, G., Wallén, P., & Grillner, S. (2021). The cyclostome spinal cord. In C. Watson, G. Sengul, and G. Paxinos (Eds.), *The mammalian spinal cord.* (pp. 213–237), Elsevier.

Robles, E., Laurell, E., & Baier, H. (2014). The retinal projectome reveals brain-area-specific visual representations generated by ganglion cell diversity. *Curr. Biol.*, *24*, 2085–2096.

Roland, P. E., Larsen, B., Lassen, N. A., & Skinhoj, E. (1980a). Supplementary motor area and other cortical areas in organization of voluntary movements in man. *J. Neurophysiol.*, *43*, 118–136.

Roland, P. E., Skinhoj, E., Lassen, N. A., & Larsen, B. (1980b). Different cortical areas in man in organization of voluntary movements in extrapersonal space. *J. Neurophysiol.*, *43*, 137–150.

Roseberry, T. K., Lee, A. M., Lalive, A. L., Wilbrecht, L., Bonci, A., & Kreitzer, A. C. (2016). Cell-type-specific control of brainstem locomotor circuits by Basal Ganglia. *Cell*, *164*, 526–537.

Rossignol, S., Dubuc, R., & Gossard, J. P. (2006). Dynamic sensorimotor interactions in locomotion. *Physiol. Rev.*, *86*, 89–154.

Rovainen, C. M. (1974). Synaptic interactions of reticulospinal neurons and nerve cells in the spinal cord of the sea lamprey. *J. Comp. Neurol.*, *154*, 207–223.

Ruder, L., Schina, R., Kanodia, H., Valencia-Garcia, S., Pivetta, C., & Arber, S. (2021). A functional map for diverse forelimb actions within brainstem circuitry. *Nature*, *590*, 445–450.

Ryczko, D., Auclair, F., Cabelguen, J. M., Dubuc, R. (2016a). The mesencephalic locomotor region sends a bilateral glutamatergic drive to hindbrain reticulospinal neurons in a tetrapod. *J. Comp. Neurol.*, *524*, 1361–1383.

Ryczko, D., Cone, J. J., Alpert, M. H., Goetz, L., Auclair, F., Dube, C., Parent, M., Roitman, M. F., Alford, S., & Dubuc, R. (2016b). A descending dopamine pathway conserved from basal vertebrates to mammals. *Proc. Natl. Acad. Sci. USA*, *113*, E2440–2449.

Ryczko, D., Gratsch, S., Alpert, M. H., Cone, J. J., Kasemir, J., Ruthe, A., Beausejour, P. A., Auclair, F., Roitman, M. F., Alford, S., & Dubuc, R. (2020). Descending dopaminergic inputs to reticulospinal neurons promote locomotor movements. *J. Neurosci.*, *40*, 8478–8490.

Sacrey, L. A., Alaverdashvili, M., & Whishaw, I. Q. (2009). Similar hand shaping in reaching-for-food (skilled reaching) in rats and humans provides evidence of homology in release, collection, and manipulation movements. *Behav. Brain Res.*, *204*, 153–161.

Sahibzada, N., Dean, P., & Redgrave, P. (1986). Movements resembling orientation or avoidance elicited by electrical stimulation of the superior colliculus in rats. *J. Neurosci.*, *6*, 723–733.

Saitoh, K., Menard, A., & Grillner, S. (2007). Tectal control of locomotion, steering, and eye movements in lamprey. *J. Neurophysiol.*, *97*, 3093–3108.

Samotus, O., Parrent, A., & Jog, M. (2018). Spinal cord stimulation therapy for gait dysfunction in advanced Parkinson's disease patients. *Mov. Disord.*, *33*, 783–792.

Santana, M. B., Halje, P., Simplicio, H., Richter, U., Freire, M. A. M., Petersson, P., Fuentes, R., & Nicolelis, M. A. L. (2014). Spinal cord stimulation alleviates motor deficits in a primate model of Parkinson disease. *Neuron*, *84*, 716–722.

Schwab, R. S., Chafetz, M. E., & Walker, S. (1954). Control of two simultaneous voluntary motor acts in normals and in parkinsonism. *AMA Arch. Neurol. Psychiatry*, *72*, 591–598.

Schwartz, A. B. (1994). Direct cortical representation of drawing. *Science*, *265*, 540–542.

Schwartz, A. B. (2007). Useful signals from motor cortex. *J. Physiol.*, *579*, 581–601.

Schwartz, A. B., Kettner, R. E., & Georgopoulos, A. P. (1988). Primate motor cortex and free arm movements to visual targets in three-dimensional space. I. Relations between single cell discharge and direction of movement. *J. Neurosci.*, *8*, 2913–2927.

Shekhar, K., & Sanes, J. R. (2021). Generating and using transcriptomically based retinal cell atlases. *Annu. Rev. Vis. Sci.*, *7*, 43–72.

Sherrington, C. S. (1906). *The integrative action of the nervous system.* Yale University Press.

Sherrington, C. S. (1924). Problems of muscular receptivity. *Nature*, *113*, 892–894.

Shik, M. L., Severin, F. V., & Orlovskii, G. N. (1966). Control of walking and running by means of electric stimulation of the midbrain. *Biofizika*, *11*, 659–666.

Slaoui Hasnaoui, M., Arsenault, I., Verdier, D., Obeid, S., & Kolta, A. (2020). Functional connectivity between the trigeminal main sensory nucleus and the trigeminal motor nucleus. *Front. Cell Neurosci.*, *14*, 167.

Smith, Y., & Bolam, J. P. (1991). Convergence of synaptic inputs from the striatum and the globus pallidus onto identified nigrocollicular cells in the rat: A double anterograde labelling study. *Neuroscience*, *44*, 45–73.

Song, J., Dahlberg, E., & El Manira, A. (2018). V2a interneuron diversity tailors spinal circuit organization to control the vigor of locomotor movements. *Nat. Commun.*, *9(1)*, 3370.

Song, J., Pallucchi, I., Ausborn, J., Ampatzis, K., Bertuzzi, M., Fontanel, P., Picton, L. D., & El Manira, A. (2020). Multiple rhythm-generating circuits act in tandem with pacemaker properties to control the start and speed of locomotion. *Neuron*, *105*, 1048–1061 e1044.

Sparks, D. L. (1986). Translation of sensory signals into commands for control of saccadic eye movements: Role of primate superior colliculus. *Physiol. Rev.*, *66*, 118–171.

Spillantini, M. G., Schmidt, M. L., Lee, V. M., Trojanowski, J. Q., Jakes, R., & Goedert, M. (1997). Alpha-synuclein in Lewy bodies. *Nature*, *388*, 839–840.

Stagkourakis, S., Spigolon, G., Liu, G., & Anderson, D. J. (2020). Experience-dependent plasticity in an innate social behavior is mediated by hypothalamic LTP. *Proc. Natl. Acad. Sci. USA*, *117*, 25789–25799.

Stein, B. E., & Stanford, T. R. (2008). Multisensory integration: current issues from the perspective of the single neuron. *Nat. Rev. Neurosci.*, *9*, 255–266.

Stein, P. S. (2008). Motor pattern deletions and modular organization of turtle spinal cord. *Brain Res. Rev.*, *57*, 118–124.

Steiner, L. A., Barreda Tomas, F. J., Planert, H., Alle, H., Vida, I., & Geiger, J. R. P. (2019). Connectivity and dynamics underlying synaptic control of the subthalamic nucleus. *J. Neurosci.*, *39*, 2470–2481.

Stephenson-Jones, M., Ericsson, J., Robertson, B., & Grillner, S. (2012a). Evolution of the basal ganglia: Dual-output pathways conserved throughout vertebrate phylogeny. *J. Comp. Neurol.*, *520*, 2957–2973.

Stephenson-Jones, M., Floros, O., Robertson, B., & Grillner, S. (2012b). Evolutionary conservation of the habenular nuclei and their circuitry controlling the dopamine and 5-hydroxytryptophan (5-HT) systems. *Proc. Natl. Acad. Sci. USA*, *109*, E164–E173.

Stephenson-Jones, M., Kardamakis, A. A., Robertson, B., & Grillner, S. (2013). Independent circuits in the basal ganglia for the evaluation and selection of actions. *Proc. Natl. Acad. Sci. USA*, *110*, E3670–E3679.

Stephenson-Jones, M., Samuelsson, E., Ericsson, J., Robertson, B., & Grillner, S. (2011). Evolutionary conservation of the basal ganglia as a common vertebrate mechanism for action selection. *Curr. Biol.*, *21*, 1081–1091.

Stephenson-Jones, M., Yu, K., Ahrens, S., Tucciaarone, J. M., van Huijstee, A. N., Mejia, L. A., Penzo, M. A., Tai, L.-H., Wilbrecht, L., & Li, B. (2016). A basal ganglia circuit for evaluating action outcomes. *Nature*, *539*, 289–293.

Stepniewska, I., Gharbawie, O. A., Burish, M. J., & Kaas, J. H. (2014). Effects of muscimol inactivations of functional domains in motor, premotor, and posterior parietal cortex on complex movements evoked by electrical stimulation. *J. Neurophysiol.*, *111*, 1100–1119.

Stepniewska, I., Pirkle, S. C., Roy, T., & Kaas, J. H. (2020). Functionally matched domains in parietal-frontal cortex of monkeys project to overlapping regions of the striatum. *Prog. Neurobiol.*, *195*, 101864.

Stepniewska, I., Pouget, P., & Kaas, J. H. (2018). Frontal eye field in prosimian galagos: Intracortical microstimulation and tracing studies. *J. Comp. Neurol.*, *526*, 626–652.

Strick, P. L., Dum, R. P., & Fiez, J. A. (2009). Cerebellum and nonmotor function. *Annu. Rev. Neurosci.*, *32*, 413–434.

Strick, P. L., Dum, R. P., & Rathelot, J. A. (2021). The cortical motor areas and the emergence of motor skills: A neuroanatomical perspective. *Annu. Rev. Neurosci.*, *44*, 425–447.

Subramanian, H. H., Arun, M., Silburn, P. A., & Holstege, G. (2016). Motor organization of positive and negative emotional vocalization in the cat midbrain periaqueductal gray. *J. Comp. Neurol.*, *524*, 1540–1557.

Subramanian, H. H., Balnave, R. J., & Holstege, G. (2021). Microstimulation in different parts of the periaqueductal gray generates different types of vocalizations in the cat. *J. Voice.* 35, 804.e9–804.e825.

Sugahara, F., Murakami, Y., Pascual-Anaya, J., & Kuratani, S. (2021a). Forebrain architecture and development in cyclostomes, with reference to the early morphology and evolution of the vertebrate head. *Brain Behav. Evol.*, 1–13.

Sugahara, F., Pascual-Anaya, J., Kuraku, S., Kuratani, S., & Murakami, Y. (2021b). Genetic mechanism for the cyclostome cerebellar neurons reveals early evolution of the vertebrate cerebellum. *Front. Cell Dev. Biol.*, *9*, 700860.

Suryanarayana, S. M., Pérez-Fernández, J., Robertson, B., & Grillner, S. (2020). The evolutionary origin of visual and somatosensory representation in the vertebrate pallium. *Nat. Ecol. Evol.*, *4*, 639–651.

Suryanarayana, S. M., Pérez-Fernández, J., Robertson, B., & Grillner, S. (2021a). The lamprey forebrain—Evolutionary implications. *Brain Behav. Evol.*, 1–16.

Suryanarayana, S. M., Pérez-Fernández, J., Robertson, B., & Grillner, S. (2021b). Olfaction in lamprey pallium revisited-dual projections of mitral and tufted cells. *Cell Rep.*, *34*, 108596.

Suryanarayana, S. M., Robertson, B., & Grillner, S. (2022). The neural bases of vertebrate motor behaviour through the lens of evolution. *Philos. Trans. R. Soc. Lond. B. Biol. Sci.*, *377*, 20200521.

Suryanarayana, S. M., Robertson, B., Wallén, P., & Grillner, S. (2017). The lamprey pallium provides a blueprint of the mammalian layered cortex. *Curr. Biol.*, *27*, 3264–3277 e3265.

Suway, S. B., & Schwartz, A. B. (2019). Activity in primary motor cortex related to visual feedback. *Cell Rep.*, *29*, 3872–3884 e3874.

Suzuki, D. G., Pérez-Fernández, J., Wibble, T., Kardamakis, A. A., & Grillner, S. (2019). The role of the optic tectum for visually evoked orienting and evasive movements. *Proc. Natl. Acad. Sci. USA*, *116*, 15272–15281.

Swanson, L. W. (2000). Cerebral hemisphere regulation of motivated behavior. *Brain Res.*, *886*, 113–164.

Takahashi, N., Oertner, T. G., Hegemann, P., & Larkum, M. E. (2016). Active cortical dendrites modulate perception. *Science*, *354*, 1587–1590.

Takakusaki, K. (2008). Forebrain control of locomotor behaviors. *Brain Res. Rev.*, *57*, 192–198.

Talpalar, A. E., Bouvier, J., Borgius, L., Fortin, G., Pierani, A., & Kiehn, O. (2013). Dual-mode operation of neuronal networks involved in left-right alternation. *Nature*, *500*, 85–88.

Tankus, A., Fried, I., & Shoham, S. (2012). Structured neuronal encoding and decoding of human speech features. *Nat. Commun.*, *3*, 1015.

Tankus, A., Lustig, Y., Fried, I., & Strauss, I. (2021). Impaired timing of speech-related neurons in the subthalamic nucleus of Parkinson disease patients suffering speech disorders. *Neurosurgery*, *89*, 800–809.

Taverna, S., Ilijic, E., & Surmeier, D. J. (2008). Recurrent collateral connections of striatal medium spiny neurons are disrupted in models of Parkinson's disease. *J. Neurosci.*, *28*, 5504–5512.

Tecuapetla, F., Jin, X., Lima, S. Q., & Costa, R. M. (2016). Complementary contributions of striatal projection pathways to action initiation and execution. *Cell*, *166*, 703–715.

Teräväinen, H., & Rovainen, C. M. (1971). Fast and slow motoneurons to body muscle of the sea lamprey. *J. Neurophysiol.*, *34*, 990–998.

Thompson, R. F., & Krupa, D. J. (1994). Organization of memory traces in the mammalian brain. *Annu. Rev. Neurosci.*, *17*, 519–549.

Thompson, R. H., Ménard, A., Pombal, M., & Grillner, S. (2008). Forebrain dopamine depletion impairs motor behavior in lamprey. *Eur. J. Neurosci.*, *27*, 1452–1460.

Tiklová, K., Björklund, A. K., Lahti, L., Fiorenzano, A., Nolbrant, S., Gillberg, L., Volakakis, N., Yokota, C., Hilscher, M. M., Hauling, T., Holmstrom, F., Joodmardi, E., Nilsson, M., Parmar, M., & Perlmann, T. (2019). Single-cell RNA sequencing reveals midbrain dopamine neuron diversity emerging during mouse brain development. *Nat. Commun.*, *10*(1), 581.

Tokuda, I. T., Hoang, H., & Kawato, M. (2017). New insights into olivo-cerebellar circuits for learning from a small training sample. *Curr. Opin. Neurobiol.*, *46*, 58–67.

Tovote, P., Esposito, M. S., Botta, P., Chaudun, F., Fadok, J. P., Markovic, M., Wolff, S. B., Ramakrishnan, C., Fenno, L., Deisseroth, K., Herry, C., Arber, S., & Luthi, A. (2016). Midbrain circuits for defensive behaviour. *Nature*, *534*, 206–212.

Travers, J. B., Dinardo, L. A., & Karimnamazi, H. (1997). Motor and premotor mechanisms of licking. *Neurosci. Biobehav. Rev.*, *21*, 631–647.

Tsutsumi, S., Chadney, O., Yiu, T. L., Baumler, E., Faraggiana, L., Beau, M., & Hausser, M. (2020). Purkinje cell activity determines the timing of sensory-evoked motor initiation. *Cell Rep.*, *33*(12), 108537.

Tye, K. M., Prakash, R., Kim, S. Y., Fenno, L. E., Grosenick, L., Zarabi, H., Thompson, K. R., Gradinaru, V., Ramakrishnan, C., & Deisseroth, K. (2011). Amygdala circuitry mediating reversible and bidirectional control of anxiety. *Nature*, *471*, 358–362.

Ullén, F., Deliagina, T. G., Orlovsky, G. N., & Grillner, S. (1997). Visual pathways for postural control and negative phototaxis in lamprey. *J. Neurophysiol.*, *78*, 960–976.

Ungerleider, L. G., &Mishkin, M. (1982). Two cortical visual systems. In D. J. Ingle, M. A. Goodale, and R. J. W. Mansfield (Eds.), *Analysis of visual behavior* (pp. 549–586). MIT Press.

Valjent, E., & Gangarossa, G. (2021). The tail of the striatum: From anatomy to connectivity and function. *Trends Neurosci.*, *44*, 203–214.

Vandaele, Y., Mahajan, N. R., Ottenheimer, D. J., Richard, J. M., Mysore, S. P., & Janak, P. H. (2019). Distinct recruitment of dorsomedial and dorsolateral striatum erodes with extended training. *Elife*, *8*, e49536.

Velliste, M., Perel, S., Spalding, M. C., Whitford, A. S., & Schwartz, A. B. (2008). Cortical control of a prosthetic arm for self-feeding. *Nature*, *453*, 1098–1101.

Vieira, E. B., Menescal-de-Oliveira, L., & Leite-Panissi, C. R. (2011). Functional mapping of the periaqueductal gray matter involved in organizing tonic immobility behavior in guinea pigs. *Behav. Brain. Res.*, *216*, 94–99.

Villalba, R. M., & Smith, Y. (2018). Loss and remodeling of striatal dendritic spines in Parkinson's disease: From homeostasis to maladaptive plasticity? *J. Neural Transm. (Vienna)*, *125*, 431–447.

von Hofsten, C., & Lindhagen, K. (1979). Observations on the development of reaching for moving objects. *J. Exp. Child Psychol.*, *28*, 158–173.

von Twickel, A., Kowatschew, D., Saltürk, M., Schauer, M., Robertson, B., Korsching, S., Walkowiak, W., Grillner, S., & Pérez-Fernández, J. (2019). Individual dopaminergic neurons of lamprey

SNc/VTA project to both the striatum and optic tectum but restrict co-release of glutamate to striatum only. *Curr. Biol.*, *29*, 677–685 e676.

Voogd, J. (1969). The importance of fiber connections in the comparative anatomy of the mammalian cerebellum. In R. Llinás (Ed.), *Neurobiology of cerebellar evolution and development* (pp. 493–514). AMA ERF Institute for Biomedical Research.

Wallén, P., & Grillner, S. (1987). N-methyl-D-aspartate receptor-induced, inherent oscillatory activity in neurons active during fictive locomotion in the lamprey. *J. Neurosci.*, *7*, 2745–2755.

Wallén, P., Robertson, B., Cangiano, L., Low, P., Bhattacharjee, A., Kaczmarek, L. K., & Grillner, S. (2007). Sodium-dependent potassium channels of a Slack-like subtype contribute to the slow afterhyperpolarization in lamprey spinal neurons. *J. Physiol.*, *585*, 75–90.

Wang, W., Schuette, P. J., Nagai, J., Tobias, B. C., Cuccovia, V. R. F. M., Ji, S., de Lima, M. A. X., La-Vu, M. Q., Maesta-Pereira, S., Chakerian, M., Leonard, S. J., Lin, L., Severino, A. L., Cahill, C. M., Canteras, N. S., Khakh, B. S., Kao, J. C., & Adhikari, A. (2021). Coordination of escape and spatial navigation circuits orchestrates versatile flight from threats. *Neuron*, *109*, 1848–1860 e1848.

Wang, Z., Kai, L., Day, M., Ronesi, J., Yin, H. H., Ding, J., Tkatch, T., Lovinger, D. M., & Surmeier, D. J. (2006). Dopaminergic control of corticostriatal long-term synaptic depression in medium spiny neurons is mediated by cholinergic interneurons. *Neuron*, *50*, 443–452.

Weglage, M., Warnberg, E., Lazaridis, I., Calvigioni, D., Tzortzi, O., & Meletis, K. (2021). Complete representation of action space and value in all dorsal striatal pathways. *Cell Rep.*, *36*, 109437.

Weiskrantz, L., Warrington, E. K., Sanders, M. D., & Marshall, J. (1974). Visual capacity in the hemianopic field following a restricted occipital ablation. *Brain*, *97*, 709–728.

Whishaw, I. Q., & Coles, B. L. (1996). Varieties of paw and digit movement during spontaneous food handling in rats: postures, bimanual coordination, preferences, and the effect of forelimb cortex lesions. *Behav. Brain Res.*, *77*, 135–148.

Whishaw, I. Q., Pellis, S. M., & Gorny, B. P. (1992). Skilled reaching in rats and humans: evidence for parallel development or homology. *Behav. Brain Res.*, *47*, 59–70.

Wibble, T., Pansell, T., Grillner, S., & Pérez-Fernández, J. (2022). Conserved subcortical processing in visuovestibular gaze control. *Nat. Commun.*, *13*, 4699.

Wiesenfeld, Z., Halpern, B. P., & Tapper, D. N. (1977). Licking behavior: Evidence of hypoglossal oscillator. *Science*, *196*, 1122–1124.

Williams, T. L., Grillner, S., Smoljaninov, V. V., Wallén, P., Kashin, S., & Rossignol, S. (1989). Locomotion in lamprey and trout: The relative timing of activation and movement. *J. Exp. Biol.*, *143*, 559–566.

Wilson, C. J., Chang, H. T., & Kitai, S. T. (1990). Firing patterns and synaptic potentials of identified giant aspiny interneurons in the rat neostriatum. *J. Neurosci.*, *10*, 508–519.

Wilson, J. J., Alexandre, N., Trentin, C., & Tripodi, M. (2018). Three-dimensional representation of motor space in the mouse superior colliculus. *Curr. Biol.*, *28*, 1744–1755 e1712.

Wolff, S. B. E., Ko, R., & Ölveczky, B. P. (2022). Distinct roles for motor cortical and thalamic inputs to striatum during motor learning and execution. *Sci. Adv.* 8(8). doi: 10.1126/sciadv.abk0231.

Wullimann, M. F., Rupp, B., & Rei.chert, H. (1996). *Neuroanatomy of the zebrafish brain: A topographic atlas*. Birkhäuser.

Yakovenko, S., & Drew, T. (2015). Similar motor cortical control mechanisms for precise limb control during reaching and locomotion. *J. Neurosci.*, *35*, 14476–14490.

Yanagihara, D., & Udo, M. (1994). Climbing fiber responses in cerebellar vermal Purkinje cells during perturbed locomotion in decerebrate cats. *Neurosci. Res.*, *19*, 245–248.

Zaccarella, E., Papitto, G., & Friederici, A. D. (2021). Language and action in Broca's area: Computational differentiation and cortical segregation. *Brain Cogn.*, *147*, 105651.

Zang, Y., Hong, S., & De Schutter, E. (2020). Firing rate-dependent phase responses of Purkinje cells support transient oscillations. *Elife*, *9*, e60692.

Zhou, F. W., Jin, Y., Matta, S. G., Xu, M., & Zhou, F. M. (2009). An ultra-short dopamine pathway regulates Basal Ganglia output. *J. Neurosci.*, *29*, 10424–10435.

Zimmerman, C. A. (2020). The origins of thirst. *Science, 370*, 45–46.

Znamenskiy, P., & Zador, A. M. (2013). Corticostriatal neurons in auditory cortex drive decisions during auditory discrimination. *Nature, 497*, 482–485.

Index

Note: Page numbers followed by *f* indicate figures, and page numbers followed by *t* indicate tables.